Ethics in Climate Change

edited by Paul Babie

T0126258

Ethics in Climate Change:

Choosing the Future

edited by Paul Babie

Adelaide
2021

Agathon: A Journal of Ethics and Value in the Modern World
Volume 8, 2021

Agathon refers to the Greek word used by Plato in the *Republic* to refer to 'the good beyond being', most notably deployed in recent times by Iris Murdoch and Emmanuel Levinas, both of whom use this touchstone to situate ethics at the heart of all of philosophy

We live in an evolving and increasingly complex world but our ethical concepts have frequently struggled to keep pace with the change. As a result, much of what passes for public debate at present remains in the grip of either deterministic or consequentialist thinking, both built on outdated assumptions and both representing attempts to address major issues in the absence of ethical concepts. We suffer, as Iris Murdoch lamented, a 'loss of concepts, the loss of a moral and political vocabulary'.

The interdisciplinary journal, *agathon*, seeks to bring together scholars from across the humanities, social sciences and sciences, including disciplines such as philosophy, theology, law and medicine, to engage with the ethical questions that now beset the modern world..

The journal is a home for considering questions such as how we deal with competing values in ethical discourse, how ethical theory finds expression in practice, what constitutes ethical character and how it is cultivated, and what excellence and wisdom look like for the ethical person or society in the twenty-first century.

Agathon is an international and interdisciplinary refereed journal published annually by ATF Press.

Chief Editor
Dr Paul Babie, University of Adelaide Law School Professor of the Theory and Law of Property

Editorial Board
- Professor Terence Lovat, Editor in Chief Bonhoeffer Legacy, The University of Newcastle, Australia & Hon Fellow University of Oxford, UK.
- Professor Robert Crotty, former Director, Ethics Centre of South Australia, Emeritus Professor of Religion and Education, University of South Australia, Adelaide.
- Managing Editor and Publisher Mr Hilary Regan, Publisher, ATF Press Publishing Group, PO Box 234 Brompton, SA 5007, Australia. Email: hdregan@atf.org.au

Subscription Rates
Local: Individual Aus $55 Institutions Aus $65 Overseas: Individuals US $60 Institutions US $65

Agathon is published by ATF Press an imprint of the ATF Press Publishing Group which is owned by ATF (Australia) Ltd (ABN 90 116 359 963) and is published once a year. ISSN 2201-3563

ISBN: 978-1-922582-32-4   soft
          978-1-922582-33-1   hard
          978-1-922582-34-8   epub
          978-1-922582-35-5   pdf

Published by:

An imprint of the ATF Press Publishing
Group owned by ATF (Australia) Ltd.
PO Box 234
Brompton, SA 5007
Australia
ABN 90 116 359 963
www.atfpress.com
Making a lasting impact

Agathon: A Journal of Ethics and Value in the Modern World, Vol 8/2021

Table of Contents

Editorial: Responding to the Risks of Climate Change   vii
     Paul Babie

Stateless, Placeless and Landless: The Complexity of Climate
     Induced Displacement/Migration in the Pacific and Its
     Implications for a Theology of Place   1
     Seforosa Carroll

Science Does Not Make Ethical Prescriptions: But in the
     Case of Climate Change Maybe it Should?   21
     Claire Williams

A Four-Step Process for Formulating and Evaluating Legal
     Commitments Under the Paris Agreement   49
     Donald A Brown, Hugh Breakey, Peter Burdon,
     Brendan Mackey, Prue Taylor

A Matter of Choice: Property and the Person   73
     Paul Babie

Emerging Perspectives on the Freedom to Choose:
     Two Pathways Out of Individualism   85
     Jana L Norman

Contributors   101

Editorial:
Responding to the Risks of Climate Change

Paul Babie

The immediate challenge of COVID-19 in 2020 and beyond notwithstanding,[1] there can be little doubt that climate change poses the most significant threat to the ongoing survival of humankind.[2] It seems that almost every day brings new evidence of this fact. The Australian summer of 2019–2020, to draw upon but one example, witnessed 'a terrible trifecta of heatwaves, drought and bushfires, made worse by climate change',[3] as depressingly summarised in this diagram:[4]

---

1. Lawrence Wright, 'The Plague Year: The Mistakes and the Struggles Behind America's Coronavirus Tragedy', in *The New Yorker* (28 December 2020); 'The Pandemic's Toll in 2020 and the Difficult Road Ahead', in *PBS News Hour* (December 31, 2020) at <https://www.pbs.org/newshour/show/the-pandemics-toll-in-2020-and-the-difficult-road-ahead>. Accessed 4 January 2021. On the need to deal with both, see Tim Flannery, *The Climate Cure: Solving the Climate Emergency in the Era of COVID-19* (Melbourne: Text, 2020).
2. Intergovernmental Panel on Climate Change (IPCC), *Global Warming of 1.5°C—Summary for Policymakers* (IPCC, 2018) at <http://www.ipcc.ch/report/sr15/>. Accessed 4 January 2021; CSIRO & Australian Government Bureau of Meteorology, *State of the Climate 2020* (Canberra: CSIRO, 2020) at <https://www.csiro.au/en/Showcase/state-of-the-climate>. Accessed 4 January 2021.
3. Climate Council, *Dangerous Summer: Escalating Bushfire, Heat and Drought Risk* (Sydney: Climate Council, 2019) at <https://www.climatecouncil.org.au/resources/dangerous-summer-escalating-bushfire-heat-drought-risk/>. Accessed 4 January 2021. See also Joëlle Gergis, *Sunburnt Country: The History and Future of Climate Change in Australia* (Melbourne: Melbourne University Press, 2018).
4. Climate Council, '2018/197 Angry Summer Infographic' at <https://www.climatecouncil.org.au/wp-content/uploads/2020/01/angry-summer-thumb.png>. Accessed 4 January 2021.

The Australian Climate Council writes that '[a] long-term warming trend from the burning of coal, oil and gas is supercharging extreme weather events, putting Australian lives, our economy and our environment at risk.'[5] The same trends are found the world over.[6] Things got no better in 2019 and 2020; at some point in the next 100 years, 2019, which is the hottest year on record, will be considered one of the cool-

---

5. Climate Council, *Dangerous Summer*.

6. 'Paris-anniversary Climate Pledges Bring Progress But Fall Short', in *The Economist* (13 December 2020) at <https://www.economist.com/international/2020/12/13/paris-anniversary-climate-pledges-bring-progress-but-fall-short>. Accessed 4 January 2021.

est.[7] Wildfires continue to burn across the Western U.S.,[8] while the 2020 Atlantic Hurricane Season is the worst on record, and is still not over.[9]

In late 2018, the Intergovernmental Panel on Climate Change (IPCC) reminded the world that climate change is now very close to the point of no return, to causing catastrophic consequences from which we may never recover. Limiting global warming to 1.5°C above pre-industrial levels would avoid many of the catastrophic consequences.[10] But beyond that, catastrophe looms. What difference does that 0.5C make? Consider this diagram:

	1.5°C	2°C	IMPACT of 2°C compared to 1.5°C
LOSS OF PLANT SPECIES	8% of plants will lose 1/2 their habitable area	16% of plants will lose 1/2 their habitable area	2x worse
LOSS OF INSECT SPECIES	6% of insects will lose 1/2 their habitable area	18% of insects will lose 1/2 their habitable area	3x worse
FURTHER DECLINE IN CORAL REEFS	70% to 90%	99%	up to 29% worse
EXTREME HEAT	14%	37%	2.6x worse
SEA-ICE-FREE SUMMERS IN THE ARCTIC	At least once every 100 years	At least once every 10 years	10x worse

CLIMATECOUNCIL.ORG.AU | crowd-funded science information

7. CSIRO & Australian Government Bureau of Meteorology, *State of the Climate*.
8. Alejandra Borunda, 'The Science Connecting Wildfires to Climate Change', in *National Geographic* (17 September 2020) at <https://www.nationalgeographic.com/science/2020/09/climate-change-increases-risk-fires-western-us/#close>. Accessed 4 January 2021.
9. Rafi Letzter, '2020 Atlantic hurricane season shatters record', in *Livescience* (10 November 2020) at <https://www.livescience.com/2020-hurricane-season-record.html>. Accessed 4 January 2021.
10. IPCC, *Global Warming of 1.5°C*.

Through our own actions, over a relatively short period of human history, we have changed the pace of geological time and, in so doing,[11] placed our own survival and that of our planet and its non-human inhabitants at grave risk.[12] We have created our own existential threat. David Wallace-Wells summarises this bleak picture with a stark warning: climate change is already, right now, '. . . worse, much worse, than you think'.[13]

The word that ought to grab our attention is 'risk'. What are the risks we are creating for ourselves and our planet, and what are we doing about them? Because while our activities are producing anthropogenic climate change, '[m]ost countries are [simultaneously] attempting to achieve environmentally and socially sustainable economic growth, coupled with food, water and energy security at a time of enormous global changes, including environmental degradation at the local, regional and global scale'.[14] The risks include:

- food, water and human security
- the economy (loss of natural capital)
- poverty alleviation and the livelihoods of the poor
- human health
- efforts to reduce the loss of biodiversity and ecosystem degradation
- personal, national and regional security[15]

To respond, we tend to see governments and corporations addressing the risks largely through strengthening the science-policy framework and, to a lesser extent, through some form of global regulatory framework,[16] such as the Paris Climate Agreement ('Paris Agreement').[17] But are these the only, or even the best ways, to respond to the risks? As to the latter, in short, while

---

11. A number of recent contributions make this point, but perhaps the most compelling is Andri Snær Magnason, *On Time and Water*, translated by Lytton Smith (London: Serpent's Tail, 2020).
12. Robert Watson, 'Risk in the Context of (Human-induced) Climate Change', in *Risk*, edited by Layla Skinns, Michael Scott and Tony Cox (Cambridge: Cambridge University Press, 2011), 158–180, 158.
13. David Wallace-Wells, *Uninhabitable Earth* (New York: Penguin, 2019), 1.
14. Watson, 'Risk in the Context of (Human-induced) Climate Change', 158.
15. Watson, 'Risk in the Context', 159.
16. Watson, 'Risk in the Context', 175–177.
17. Conference of the Parties, United Nations Framework Convention on Climate Change, *Report of the Conference of the Parties on Its Twenty-First Session*, Held in Paris from 30 November to 13 December 2015, FCCC/CP/2015/L.9/Re.1.

> [a] suitable policy framework would facilitate the emergence of appropriate pricing and technological mechanisms[] [a] voluntary agreement will not work. Instead, we need a long-term (e.g. 2030–2050), legally binding global regulatory framework that involves all major emitters[,] . . . allocat[ing] responsibilities in an equitable manner . . . and should include immediate and intermediate targets that are differentiated.[18]

This may seem rather unlikely from our current vantage point. Global governance proves particularly vexing to those committed to the present state of our world, such as those who, as does much of the globe, see the nation state structured around capitalism and democracy as the only way in which to organise our political, social, and economic lives.[19] But more significantly might be the way in which we think of climate change itself:

> The dangers of climate change are not somehow 'out there', like a danger of alien invasion—external to the way we live and organise ourselves. The dangers of climate change are 'in here'—a function of human technologies, social relationships, economic and political systems. This places a limit on what science can achieve . . . Yes, the risks associated with a changing climate are mediated by the physical processes of the atmosphere, oceans, ice sheets and biosphere. And science is well placed to help us understand how these processes work. But the climate risks we imagine or experience are also powerfully mediated by the ways we live and argue together. This means that to claim to be able to 'fix' climate change— in the sense of reducing or eliminating weather risks and the associated dangers of climate change—would be to claim to 'fix' society in some way. Any purely technological claim to fix climate change . . . misses at least half the story.[20]

Climate change, in truth, as many say, is a challenge. What that really means is that we are challenged, of course, to find scientific and regulatory tools that will help in addressing it. But those will not solve the problem. What we are really being challenged to do is to change the

---

18. Watson, 'Risk in the Context of (Human-induced) Climate Change', 176.
19. On this, see David Shearman and Joseph Wayne Smith, *The Climate Change Challenge and the Failure of Democracy* (Westport, CT: Praeger, 2007).
20. Mike Hulme, *Can Science Fix Climate Change: A Case Against Climate Engineering* (Cambridge, UK: Polity, 2014), 130–131.

way we live, to think of new ways of organising our collective lives. Simply, climate change is not only a challenge for scientists and the formulators of law and policy, nationally or globally. Instead, it is a challenge to develop a new moral order,[21] one that is founded upon pluralism and pragmatism.[22] The former 'offers checks and balances to scientific and political projects alike. It recognises the inevitability of competing values and goals and concedes that a world of more than 7 billion people cannot move together.'[23]

> The corollary of pluralism is philosophical and political pragmatism[, which] foregoes the goal of establishing an ultimate truth in favour of working with merely satisfactory truths, while political pragmatism suggests a cautious and flexible approach to defining the problems we decide to tackle. Pragmatism is thus content to recognise and name problems like climate change as being super-wicked in character: not definable and solvable. Instead of using science and technology to 'fix' wicked problems, pragmatism is content to pursue multiple and clumsy solutions to regularly reframed problems in order to achieve merely incremental gains.[24]

This special issue seeks this path of pragmatism, incrementally searching out a new moral order, a new way of living collectively that recognises the risks, and responds to them in humility. In five contributions, we seek an outline of what is necessary if we are to change the way we live.

We begin, though, with the legal and policy framework, and with science, for a simple reason: whatever more we must do to seek this incremental path towards a new morality, we must begin with what is required by the legal obligations undertaken by the parties to the Paris Agreement. And to understand what that might be, we do need to begin with the science. As noted here and throughout the contributions to this special issue, science plays an important role in telling

---

21. Mike Hulme, 'The True Meaning of Climate Change', in *New Scientist* (5 September 2009), 28–29; Mike Hulme, *Why We Disagree About Climate Change: Understanding Controversy, Inaction and Opportunity* (Cambridge: Cambridge University Press, 2009), 362.
22. Hulme, *Can Science Fix Climate Change*, 137–40.
23. Hulme, *Can Science Fix Climate Change*, 137.
24. Hulme, *Can Science Fix Climate Change*, 137–138.

us about how climate change works and what it is doing to our planet. Moreover, there is near complete consensus about that science.

In 'Stateless, Placeless, and Landless: the complexity of climate induced displacement in the Pacific', Seforosa Carroll explores the challenges faced by Pacific Islanders through climate displacement and migration, viewed through the lens of Christian theology. The Pacific Island Countries (PICs) have been identified as a cluster of countries under threat due to climate change. For low-lying atoll countries external migration looms large as internal relocation is limited. For countries like Kiribati and Tuvalu migration in the form of forced relocation is an imminent option that they will need to consider. The importance of the imminent plight of countries like Tuvalu and Kirbati raise a number of complex theological questions and conceptual problems in relation to land, state and place. Forced relocation or migration involves a number of issues that will need to be addressed, including the preservation of identity and culture of a community as well as the role of international law, theology, church and pastoral practice. Receiving countries, too must prepare for education and awareness; it is estimated that 250 million people worldwide will be displaced by environmental and climatic changes by 2050. Carroll argues that the problem of the 'disappearing islands' in the Pacific is that it assumes that the submergence and eventual disappearance of land equates with the loss of country and people. Carroll suggests this is not likely the case; instead, the most probable likelihood is that the land will become uninhabitable before it disappears. People therefore face the paradox of becoming 'stateless' before the actual physical disappearance of their country. Carroll, then, explores the challenges for a theology of place by (re)considering the meaning of place when there is no longer a land/country to return to (the absence of a physical home) and the concomitant statelessness which follows.

Carroll points us towards another question. If the science of climate change shows us what is happening to everyone on this planet, as represented most alarmingly in the Pacific Islands, what ought we to do to respond to that reality? Science fails to offer the only answers, but it can still offer part of the solution. Claire Williams, in 'Science Does Not Make Ethical Prescriptions—But in the Case of Climate Change Maybe It Should?', shows how, arguing that law, as the over-arching institution governing our behaviour, ought in some way to reflect scientific knowledge about the human impact on Earth. While this may seem a logical proposition on the surface, it brings up complex ethical and practical questions regarding the role of pure scientific inquiry, and in particular the obligation, if any, that science ought to play in shaping law and policy. Williams examines the moral difficulties in 'reading' from science how we should conduct our communities and ourselves, as well as the inherent problems with deriving an 'ought' from an 'is,' often referred to as 'Hume's gap,' after 18th century philosopher David Hume. For science does not make ethical prescriptions. But in the case of climate change maybe it should.

From the scientific foundations for what law and policy ought to do, we turn to the international legal obligations. In 'A Four-Step Process for Formulating and Evaluating Legal Commitments Under the Paris Agreement', Donald A Brown, Hugh Breakey, Peter Burdon, Brendan Mackey, and Prue Taylor attempt to work out what exactly will be required of nations by the Paris Agreement to respond to the challenge. The Paris Agreement requires each party to prepare, communicate and maintain successive Nationally Determined Contributions (NDCs) that it intends to pursue through domestic mitigation measures. NDCs seek a reduction of greenhouse gas (GHG) emissions so as to hold the increase in the global average temperature to well below 2°C above pre-industrial levels, and to pursue efforts to limit the increase to 1.5°C above pre-industrial levels. In accepting the Paris Agreement, Brown, Breakey, Burdon, Mackey, and Taylor argue that nations have accepted both legally binding non-discretionary duties *and* normative obligations about which nations may exercise some discretion and, in doing so, make explicit the moral (or ethical) responsibilities implicit in those obligations. The four steps that all nations should therefore expressly consider in formulating an NDC are: (1) select a global warming limit to be achieved by the GHG emissions reduction target; (2) identify a global carbon budget consistent with achieving the global warming limit at an acceptable level of probability; (3) determine the national fair share of the global carbon budget, based upon equity and common but differentiated responsibilities and respective capabilities; and (4) specify the annual rate of its national GHG emissions reductions on the pathway to net zero emissions. These steps serve as a guide for nations when developing and communicating their NDCs. The process also provides guidance for other stakeholders (for example, non-governmental organisations, states and others in the international community) through formal processes such as the Global Stocktake, who seek to evaluate the level of ambition and fairness entailed by NDCs. Nations are of course free to formulate their NDCs after consideration of issues that go beyond the global carbon budget and equity considerations, such as, for instance, the obligations of developed nations to assist developing nations with GHG emissions reduction commitments.

From international moral (ethical) obligations imposed on nations, I turn to those which might be asked of individuals living within those nations. In 'A Matter of Choice: Property and the Person', I argue that the liberal conception of private property is not a solution to anthropogenic climate change but, rather, the source of the problem. The concept of private property facilitates the human activities that cause anthropogenic climate change and the resulting human externalities suffered disproportionately by those in the developing world. The essay concludes that we can no longer wait for government to act; we must take individual, personal, action now if we are to address the challenge of climate change. That action involves limiting our own exercise of choice in dealing with those things said to be our property. Everything that is ours has, in one way or another, the potential to create GHG emissions, either in the way it was produced before we owned it, or in the way we use it once we do own it. In both cases, our responsibility is to exercise choice so as to own goods and resources that themselves are created through a reduction in emissions as well as to use those things that we do own in ways that limit emissions. The question of property, I argue, is one of choice—we have the choice to modify our own lifestyles so as to reduce our climate footprint.

How, though, do we exercise choice itself? This is really a question of the nature of the individual, the self. Jana Norman addresses that question in 'Emerging Perspectives on the Freedom to Choose: Two Pathways Out of Individualism', in which she argues that given the near total consensus (97 percent) amongst climate scientists that climate change is extremely likely due to human activities, the significant percentages of people professing not to *believe in* anthropogenic climate change seems, in and of itself, incredible. This, Norman argues, is the territory of psychology, sociology, and anthropology, and a debate over culture, worldviews, and ideology. The climate crisis is, in this sense, an existential crisis, forcing us to change the way we see the world and live within it. And no matter how illogical it seems in terms of the science, it makes sense that denial is one of the responses to this existential crisis. Denial is always part of human response to the upheaval, grief and loss we experience when our world changes either interpersonally or globally, as in the case of climate. For Norman, 'a better story about who we are and our place in the world as human beings would go a long way towards helping us get our bearings in this reality, helping some move out of denial and ameliorating some of the adverse mental health effects for those of any political stripe who struggle with how to live in this reality and face its consequential challenges to our sense of identity, belonging and purpose in the world.' Norman presents two projects—one from a Christian perspective and one from the perspective of the philosophy of science—which can be characterised as the twin pillars of the western social imaginary. These stories neither diminish the self nor limit choice, but offer a corrective to individualism, reframing the value of freedom as the freedom to choose life.

Each of these contributions takes a plural and pragmatic approach. There are no sweeping claims about our solutions providing the ultimate panacea. Instead, in each case, humility informs the proposal— clear that what is presented here is done so not as the ultimate fix, but as an incremental offering of what might be tried as we, humanity, respond to the challenge of climate change. None of the sciences or the social sciences and humanities, alone, can fix this problem. Indeed, it may not be possible to 'fix' this problem. What each discipline can do, however, is offer, in humility, a suggestion that will make a small change to be added to many other small changes. The hope, of course, is that these incremental changes will produce a new way of living on this planet, one characterised by humility, pluralism, and pragmatism.

Agathon: A Journal of Ethics and Value in the Modern World, Vol 8/2021

Stateless, Placeless and Landless: The Complexity of Climate Induced Displacement/Migration in the Pacific and Its Implications for a Theology of Place

Seforosa Carroll

Abstract: The Pacific Island Countries (PICs) have been identified as a cluster of countries under threat due to climate change. For low-lying atoll countries external migration looms large as internal relocation is limited. For countries like Kiribati and Tuvalu migration in the form of forced relocation is an imminent option that they will need to consider. The importance of the imminent plight of countries like Tuvalu and Kiribati raise a number of complex theological questions and conceptual problems in relation to land, state and place that this paper seeks to chart. Forced relocation or migration is not as simple as 'packing your home on your back'. It will involve a number of issues that will need to be addressed such as the preservation of identity and culture of a community as well as the role of international law, theology, church and pastoral practice. It will also need to take into account the preparation, education and awareness that will be required of receiving countries of climate induced displaced persons. It is estimated that 250 million people worldwide will be displaced by environmental and climactic changes by 2050. This paper seeks to explore forced displacement in the context of 'disappearing islands' in the Pacific. The problem with the 'disappearing islands' rhetoric is that it assumes that the submergence and eventual disappearance of land equates with the loss of country and people. But this is not likely the case. The most probable likelihood as demonstrated already by many Pacific island states today, is that the land will become uninhabitable before it disappears. People are therefore 'stateless' before the actual physical disappearance of their country. This paper explores the challengers to a theology of place by (re)considering the meaning of place when there is no longer a land/country to return to (the absence of a physical home) and statelessness.

The impact of climate change is anticipated to displace up to 250 million people worldwide by 2050.[1] The office of the United Nations High Commissioner for Refugees (UNHCR) estimates that 'an annual average of 21.5 million people have been forcibly displaced by weather-related sudden onset hazards—such as floods, storms, wildfires, extreme temperature—each year since 2008'.[2] The Internal Displacement Monitoring Center's (IDMC) 2016 global displacement report recorded 19.2 million NEW displacements across 113 countries as a result of disasters in 2015.[3] In the first six months of 2020, the IDMC reported 14.6 million new internal displacements across 127 countries of which 9.8 million was due to disasters and the remaining 4.8 million triggered by conflict and violence.[4]

The Pacific, 'the liquid continent', is among a number of countries in the world on the forefront of climate change. The Pacific Island Countries (PICs) have been identified as a cluster of countries under threat due to climate change 'in international agreements, academic writings and the media.[5] The Pacific region has been identified as one

1. See Gulrez Shah Azhar, 'Climate change will displace millions in coming decades: Nations should prepare now to help them', in *The Conversation* (19 December 2017) at <https://theconversation.com/climate-change-will-displace-millions-in-coming-decades-nations-should-prepare-now-to-help-them-89274>. Accessed 30 January 2021; Melita Sunjic, 'Top UNHCR official warns about displacement from climate change', in *United Nations High Commissioner for Refugees* (9 December 2008) at <https://www.unhcr.org/news/latest/2008/12/493e9bd94/top-unhcr-official-warns-displacement-climate-change.html>. Accessed 30 January 2021.
2. Internal Displacement Monitoring Centre, *Global Report on Internal Displacement (GRID) 2016* (Geneva, Switzerland: Internal Displacement Monitoring Centre, 2016) available at <https://www.internal-displacement.org/globalreport2016/#home>. Accessed 30 June 2020.
3. (Geneva, Switzerland: Internal Displacement Monitoring Centre, 2016) available at <https://www.internal-displacement.org/globalreport2016/#home>. Accessed 30 June 2020.
4. Internal Displacement Monitoring Centre, *Internal Displacement 2020: Mid Year Update* (Geneva, Switzerland: Internal Displacement Monitoring Centre, 2020) availale at <https://www.internal-displacement.org/sites/default/files/publications/documents/2020%20Mid-year%20update.pdf>. Accessed 30 November 2020.
5. Jon Barnett and John R Campbell, *Climate Change and Small Island States: Power, Knowledge and the South Pacific* (London, UK: Earthscan, 2010).

of the world's climate change hot spots.[6] Widespread consensus sup-ported by over twenty years of research and policy recognises 'that climate change is extremely dangerous for small island developing states'.[7] The 'most significant effects of climate change include reduc-tions in agricultural productivity; reductions in water quantity and quality, with associated impacts on agriculture, health; increases in climatic events; coastal erosion and inundation as a result of extreme events and sea level rise'.[8]

The impact of climate change is already evident and felt in the Pacific in a variety of ways. For those in Tuvalu, Kiribati, Marshall Islands, Tokelau and the Maldives time is already running out. The rising sea level, the 'overflowing ocean' is drowning them out.[9] Sev-eral coastal communities across the Pacific have already been relo-cated due to environmental degradation. In Fiji for example, up to forty coastal villagers have been identified to be relocated inland due to rising sea or river levels.[10] In the Solomon Islands five islands have already been lost to the rising sea.[11] The people of the Carteret islands in Papua New Guinea have already experienced the complexities of

6. Volker Boege, 'Climate Change and Conflict in Oceania: Challengers, Responses and Suggestions for a Policy-Relevant Research Agenda', in *Toda Peace Institute Policy Brief*, 17 (2018): 2.

7. Barnett and Campbell, *Climate Change and Small Island States: Power, Knowledge and the South Pacific*, 1.

8. John R Campbell, 'Climate-Induced Community Relocation in the Pacific: The Meaning and Importance of Land', in *Climate change and displacement: multidiscilpinary perspectives*, edited by Jane McAdam (Oxford, UK: Hart, 2012), 64–65.

9. Orrin H Pilkey *et al*, editors, *Retreat From a Rising Sea: Hard Choices in An Age of Climate Change* (New York, NY: Columbia University Press, 2016).

10. See 'Fiji villagers face relocation as sea levels rise', in *ABC News* (16 August 2013) at <http://www.abc.net.au/news/2013-08-16/fiji-villages-face-relocation-as-sea-levels-rise/4891542>. Accessed 30 January 2021; 'In Fiji, villagers need to move due to climate change', in *Sydney Morning Herald* (17 November 2017) at <https://www.smh.com.au/world/fiji-to-move-more-than-40-villages-inland-as-seas-rise-20171117-gznzrx.html>. Accessed 30 January 2021; Loes Witschge, 'In Fiji, villagers need to move due to climate change', in *Aljazeera* (14 February 2018) at <https://www.aljazeera.com/indepth/features/fiji-villages-move-due-climate-change-180213155519717.html>. Accessed 30 January 2021.

11. Simon Albert et al, 'Interactions between sea level rise and wave exposure on reef island dynamics in the Solomon Islands', in 11 *Environmental Research Letters* (2016).

resettlement in Bougainville. Within the last four years Tonga, Vanuatu, Fiji and Samoa have experienced destructive category 4 and 5 cyclones. The effects of El Nino are currently being experienced in the highlands of PNG, parts of Vanuatu and Fiji. Many continue to die of hunger due to famine. It is expected that 4.37 million people in the Pacific are likely to be affected and at risk from drought.[12] The impact and effects of climate change continue to challenge and pressure island economies, habitats, and the livelihoods of people in the region giving rise to a broad spectrum of newly arising economic, social and cultural problems.[13]

In *Retreat from a rising sea: hard choices in an age of climate change*, Pilkey traces the causes behind rising sea levels and its implications for diverse coastlines.[14] The significance of the study by Pilkey *et al* is that it demonstrates the impact of global warming and rising sea level beyond the global South. Pilkey details specific threats faced by Miami, New Orleans, New York, and Amsterdam. Aware that sea level has changed throughout the earth's long history, they emphasise how global variations in the sea level 'are due to changes in the volume of the earth's oceans . . . and very gradual changes in the capacity of ocean basins' as well as the 'direction and intensity of ocean currents'.[15] They accept the 'consensus among climatologists, glaciologists and oceanographers . . . that our society should be prepared for a three foot (0.91 metres) rise by 2100.' The primary reasons for this process of rising sea levels are melting ice sheets and warming oceans and the role of human agency in the production of greenhouse gasses caused by the burning of fossil fuels. They estimate that by the end of this century, hundreds of millions of people living at low elevations along coasts will be forced to retreat to higher and safer ground. The problem, however, for low-lying islands as is the case in of Tuvalu or Kiribati, is that their people will not be able to relocate forever within their own boundaries.

---

12. United Nations Office for the Coordination of Humanitarian Affairs (UNOCHA), 'El-Niño in the Pacific', in UNOCHA (November 2015) at <https://www.humanitarianresponse.info/sites/www.humanitarianresponse.info/files/documents/files/rop_hb_nov18.pdf>. Accessed 30 January 2021.
13. Boege, 'Climate Change and Conflict in Oceania: Challengers, Responses and Suggestions for a Policy-Relevant Research Agenda', 1.
14. Pilkey *et al*, *Retreat From a Rising Sea: Hard Choices in An Age of Climate Change*.
15. Pilkey *et al*, *Retreat From a Rising Sea: Hard Choices in An Age of Climate Change*, 2.

For low-lying atoll countries external migration looms large as internal relocation is limited. For countries like Kiribati and Tuvalu migration in the form of forced relocation is an imminent option that they will need to consider. The importance of the imminent plight of countries like Tuvalu and Kiribati raise a number of complex theological questions and conceptual problems in relation to land, state and place that this article seeks to chart. Forced relocation or migration is not as simple as 'packing your home on your back' or securing a territory. It will involve a number of issues that will need to be addressed such as the preservation of sovereignty, identity and culture of a community as well as the role of international law, theology, church and pastoral practice. It will also need to take into account the preparation, education, development of culturally sensitive policies and awareness that will be required of receiving or host countries of climate induced displaced persons.

This article seeks to explore forced displacement in the context of 'disappearing islands' in the Pacific by outlining the complexities of climate induced displacement and the challenges this poses to developing a theology of place in the Pacific context, particularly where there is the likelihood of the loss of a physical home or place, as in the case of Tuvalu and Kiribati. The objective is to identify pertinent themes for a theology of place in the context of 'disappearing islands' as well as argue for the importance of the place of theology in conversations on relocation and public policy, particularly pertaining to the Pacific.

Climate induced migration in the form of displacement, relocation and resettlement will continue to compel or drive movements of people across and beyond the Pacific. For some like Tuvalu and Kiribati, forced relocation could mean the permanent loss and eternal absence of a physical home. In the absence of a physical home, how can home or place be reimagined and culture preserved? The answer to these questions will come from a combination of sources and perspectives such as international law, migration and cultural studies, climate and environmental science, history, geography, indigenous epistemologies and theology, to name a few. Although, more than often, theological or faith perspectives and indigenous epistemologies are excluded from the conversation and yet both perspectives are critical. A multi, interdisciplinary wholistic approach is required to facilitate the process of relocation aimed at supporting people to move with dignity and pride intact. The key to relocation, is careful, well thought out preparation at both the macro and micro levels and must include and take into account the agency and involvement of the people of the communities relocating. As Tammy Tabe asserts relocation 'must be carefully planned and conducted properly with consent of the communities and people. The relocation must be planned and facilitated in a way that enables the continuity and sustainability of the community in the new destinations.[16] Further, Tabe adds there are lessons that could be learned and implemented from past relocations in the Pacific.

Pacific islanders have a long history of migration.[17] Mobility, movement, journey and voyaging have always been a part of Pacific people's history. Migration, place making and cultural identifications are not new concepts for Pacific Islanders.[18] Jione Havea maintains that metaphors of voyage and navigation continue to live on in the names people *Vaka* (boat), *Moana* (ocean), and in cultural dances— *haka* (movements of the body).[19] Islanders have moved for various reasons. In colonial times community movements, some of which have been forced relocations associated with the economic and colonial interests of the Colonial Empires.[20] Examples of these kinds of movements are the relocation of the population of Banaba to Rabi island in Fiji in 1945 by the Gilbert and Ellice Islands Colony for the mining of phosphate on Banaba island. In 1946 the inhabitants of Bikini atoll were moved to Rongerik in the Marshall Islands so that the United States in America could use the atoll for nuclear testing.[21]

---

16. Tammy Tabe, 'Climate Change Migration and Displacement: Learning from Past Relocations in the Pacific', in *Social Sciences*, 8/218 (2019): 15.
17. See Campbell, 'Climate-Induced Community Relocation in the Pacific: The Meaning and Importance of Land'.
18. Jione Havea, 'The Politics of Climate Change: A Talanoa from Oceania', in *International Journal of Public Theology*, 4/3 (2010): 345; Elfriede Hermann et al, editors, *Belonging in Oceania: Movement, Place-Making and Multiple Identifications* (New York, NY, and Oxford, UK: Berghahn, 2014).
19. Jione Havea, 'Migration and Mission Routes/Roots Oceania', in *Christianities in Migration: The Global Perspective*, edited by Elaine Padilla and Peter C Phan (New York, NY: Palgrave Macmillan, 2016), 113–114.
20. Tabe, 'Climate Change Migration and Displacement: Learning from Past Relocations in the Pacific'.
21. Tabe, 'Climate Change Migration and Displacement: Learning from Past Relocations in the Pacific'.

Natural hazards have also been factors in the movements and subsequent relocation of communities. The impact of climate change experienced through sea level rise, floods, droughts, storm surges, and increasing intensity in tropical cyclones have led to more and more communities being displaced and the increasing relocation. It is perhaps the forced displacement brought about by climate change that is the most problematic for islanders. Pacific Islanders feel a deep sense of injustice. They had very little to do to cause the predicament they find now themselves in. Their foremost concern is calling polluting countries to account for their actions and to reduce their emissions and/or contribute to finding ways (and paying for costs) for adapting to the effects of climate change. Underlying this overwhelming concern is the hope that if emission levels are kept to a safe minimum, then there is the possibility that Pacific Island countries may have a longer time and opportunity to save their homes. This, however, does not address the inevitable likelihood of future relocation.

Environmental migration is a not a new phenomenon as 'changing environmental conditions have been migration drivers throughout history'.[22] What is new about this 'newly perceived form of migration' is 'recognising anthropogenic drivers of climate change which induce migration'.[23] As Scott Leckie *et al* contend 'the global displacement crisis is an outcome of decades of political and ecological displacement by the world's most polluting nations'.[24] Recognising the role developed countries have played in relation to anthropogenic climate change, the World People's Conference on Climate Change and the Rights of Mother Earth held in Cochabamba on 22 April 2010, People's Agreement states:[25]

22. László J Kulcsár, 'The Day after Tomorrow: Migration and Climate Change', in *Routledge International Handbook of Migration Studies*, edited by Steven J Gold and Stephanie J Nawyn (London, UK, and New York, NY: Routledge, 2013), 28.
23. Roger Zetter, 'Protecting People Displaced by Climate Change: Some Conceptual Changes', in *Climate Change and Displacement*, edited by Jane McAdam (Oxford, UK, and Portland, OR: Hart, 2012), 131–132.
24. Scott Leckie *et al*, editors, *Climate Change and Displacement Reader* (London, UK: Earthscan, 2012), 3.
25. *World People's Conference on Climate Change and the Rights of Mother Earth: Building the People's World Movement for Mother Earth* (2010) at <https://pwccc. wordpress.com/support/>. Accessed 30 January 2021.

> Developed countries, as the main cause of climate change, in assuming their historical responsibility, must recognize and honour their climate debt in all of its dimensions as the basis for a just, effective, and scientific solution to climate change.

It follows then as Sujatha Byravan and Sudhir Chella Rajan convincingly argue that the responsibility for accommodating those who will inevitably be displaced—'climate exiles'—by climate change impacts should be a shared global responsibility. By this they mean that the responsibility of absorbing 'climate exiles' should be shared among host countries and proportional to each country's 'cumulative emissions of greenhouse gases'.[26] They contend that to 'ignore potential victims until after they become "environmental refugees"—is morally indefensible as well as impractical.'[27]

For small low-lying atolls like Tuvalu and Kiribati whose islands are already vulnerable and at risk of being swallowed up by rising sea levels, the complexity of climate displacement and relocation is manifold. The notion of 'disappearing islands' or 'sinking islands' 'is premised on the assumption that at some point, the territories of states such as Kiribati and Tuvalu will disappear—either completely, or to the point that they can no longer sustain permanent populations'.[28] On the one hand there are implications to do with statehood and stateless and contending with the impending loss of place and land. On the other, are the emerging themes for developing a theology of place in the context of being displaced and uprooted which must take into account the agency of the communities involved, their indigenous epistemologies, as well as considerations for the pastoral, ethical and theological implications both within the Pacific context itself and beyond that must accompany the process. The notion of being stateless, placeless or landless have not only to do with the law, in the Pacific context, they strike at the heart of what it means to be a people whose identity, dignity and humanity is intricately intertwined with and to land or place. The Pacific understanding of land holds deep connections and roots that ground culture, identity, kinship, and

---

26. Byravan, and Rajan. 'The Ethical Implications of Sea-Level Rise', in *Ethical & International Affairs*, 24/3 (2010): 239, 241.
27. Byravan, and Rajan. 'The Ethical Implications of Sea-Level Rise', 242.
28. Jane McAdam, '"Disappearing States", Statelessness and the Boundaries of International Law', in *Climate Change and Displacement: Multidisciplinary Perspectives*, edited by Jane McAdam (Oxford, UK: Hart, 2010), 109.

spirituality. Articulating the devastating impact of what the loss of land and place will mean for Tuvuluans, Reverend Tafue Lusama, former General Secretary of the Ekalesia Kelisiano Tuvalu (EKT) states 'If this small space (Tuvalu) is submerged under water, that is the end, and that is the literal death of a people, of us as a people.'[29]

In the Pacific it is difficult to conceive culture, land and state apart from religious claims when embodied in so many constitutions—the *Vanua* (land), *Lotu* (church) and *Matanitu* (government). Pacific people have traditionally had a wholistic and relational worldview.[30] Leslie Boseto articulates this relationship the following way 'Our existence and our survival can never be separated from our land and sea.'[31] Being human from a Pacific perspective is understood as being intrinsically interconnected with the land, the sea, creation, community and faith. For Pacific people, land—*vanua, fenua, whenua*—encompasses and embodies the physical, spiritual, social, and cultural. As Boseto explains 'God's location is in creation. Our life in God is in creation.'[32] Within this web of interrelationships there is an understanding of how everything is interconnected; the role of human beings, the seasons, and all living creatures and plants play in keeping the delicate balance of life.[33] As Upolu Vaai asserts 'Knowledge is not separate from the land. The loss of land has a direct implication for

---

29. 'Tuvalu: Faith in a changing climate' (Australia: Market Lane Media, 2018) at <https://vimeo.com/239577029>. Accessed 30 January 2021.
30. See Upolu Lumā Vaai and Aisake Casimira, eds, *Relational Hermeneutics: Decolonising the Mindset and the Pacific Itulagi* (Suva, Fiji: The University of the South Pacific & The Pacific Theological Collge, 2017); Unaisi Nabobo-Baba, *Knowing and Learning: An Indigenous Fijian Approach* (Suva, Fiji: Institute of Pacific Studies University of the South Pacific, 2006); Havea, 'The Politics of Climate Change: A Talanoa from Oceania'.
31. Leslie Boseto, 'Do Not Separate us from Our Land and Sea', in *Pacific Journal of Theology*, 2/13 (1995): 69.
32. Leslie Boseto, 'Do Not Separate us from Our Land and Sea', 69.
33. For Pacific perspectives of the interconnections and interrelationships between creation, environment, and people, see IS Tuwere, *Vanua: Towards a Fijian Theology of Place* (Suva, Fiji: Institute of Pacific Studies, University of South Pacific and College of St John the Evangelist, 2002); Nabobo-Baba, *Knowing and Learning: An Indigenous Fijian Approach*; Rosiani Lagi, 'Vanua Sauvi: Social roles, sustainability and resilience', in *Relational Hermeneutics: Decolonising the Mindset and the Pacific Itulagi*, edited by Upolu Luma Vaai and Aisake Casimira (Suva, Fiji: The University of the South Pacific Press & The Pacific Theological College, 2017), 187–197.

the loss of cultures and language'.[34] Writing about this connection in relation to relocation, Tuvaluan Maina Talia stresses 'our special bond with our land and sea cannot be dismantled, and this what makes it difficult for us to leave Tuvalu.[35] Low lying atoll islands like Tuvalu and Kiribati are facing the challenge of impending loss of land (statehood), status as a consequence (stateless), and a sense of place. In a context where the looming horizon points a community to being stateless, placeless and landless, how would a theology of place need to be (re)framed?

The problem with the 'disappearing islands' rhetoric is that it assumes that the submergence and eventual disappearance of land equates with the loss of country and people. But this is not likely the case. The most probable likelihood as demonstrated already by several Pacific island states today, is that the land will become uninhabitable before it disappears. The four elements of statehood as contained in Article 1 of the Montevideo Convention on the Rights and Duties of States are: a defined territory, a permanent population, an effective government, and the capacity to enter into relations with other countries. In the case of Tuvalu and Kiribati 'it is more probable that the other indicia of statehood—a permanent population, an effective government and the capacity to enter into relations with other states– will have been challenged' prior to the actual disappearance of the physical land territory.[36]

Tuvalu, and other low lying atoll small island states are already at risk of losing two fundamental criteria of statehood because of climate change. The first is that of a permanent population. The effects of climate change experienced with frequent and increasing severe weather patterns such as sea level rise, tropical cyclones, droughts, floods, coastal erosions, storm surges, king tides and salt water intrusion will make the land uninhabitable. This will lead to the eventual movement of people elsewhere. In such an event the criterion for a permanent population will be lost. In the long term the submersion

---

34. Upolu Lumā Vaai, '"We are therefore we live" Pacific Eco-Relational Spirituality and Changing the Climate Change Story', in *Toda Peace Institute Policy Brief*, 56 (2019): 5.

35. Maina Talia, *Migration is a definite "No" but Rather a Matter of Choice: Voice from the Margin* (Berlin, Germany: Green Party, German Parliament, 2019), 11.

36. McAdam, '"Disappearing States", Statelessness and the Boundaries of International Law', 108.

of the physical land territory, although a gradual process, will result in the loss of the requirement of territory. The critical question then as Tuvuluan lawyer Simone Kofe asks: can Tuvalu continue to exist in the absence of a permanent population and territory?[37]

There is another added complexity. The law has never had to deal with the potential extinction of a country because of physical disappearance.[38] In the case of Kiribati and Tuvalu, people are 'stateless' before the actual physical disappearance of their country. However, this does not render them 'stateless' under article 1 of the 1954 convention relating to the Status of Stateless Persons and, subsequently in 1961, the Convention on the Reduction of Statelessness—because the definition 'is deliberately restricted to people who are 'not considered as a national by any State under the operation of its law'.[39] This means that unless people have been deprived of a nationality from that country, signatory countries are not obligated to take them in. Although as McAdam suggests, there is a small window of opportunity through the UNHCR institutional mandate. The Final Act of the Convention on the Reduction of Statelessness allows for de facto stateless persons (of which 'disappearing islands' could qualify) to be afforded the same treatment as de jure stateless persons, however, the convention 'only binds states that have ratified it, and only in relation to stateless persons within their territory'.[40] Given the limitations of international law in this regard, McAdam suggests that *en masse* migration might be the best option should in situ adaption no longer be possible. But she cautions 'if *en masse* relocation to another country is to be considered as a permanent solution, then issues other than land alone need to be considered in order to provide security for the future'.[41]

---

37. Although Kofe poses and responds to the question from a legal perspective, the question also provokes an imagining of place that moves beyond the bounds of international law. What does it mean to envision a theology of place in the absence of place? Simon Kofe, *The Legal Implications of Climate Change on the Statehood of Tuvalu* (Masters of Law) (Malta: International Maritime Law Institute, 2014).

38. McAdam, "'Disappearing States", Statelessness and the Boundaries of International Law'.

39. McAdam, "'Disappearing States", Statelessness and the Boundaries of International Law'.

40. McAdam, "'Disappearing States", Statelessness and the Boundaries of International Law'. 121.

41. McAdam, "'Disappearing States", Statelessness and the Boundaries of International Law', 123.

Relocation and resettlement are not the first option for the currently vulnerable 'sinking islands'. There are a number of reasons for this. There is a strong opposition to the label 'refugee'. The label is seen as 'invoking a sense of helplessness and lack of dignity which contradicts the very strong sense of Pacific pride'.[42] Also, the term 'refugees' implies fleeing from persecution. For people from Tuvalu and Kiribati, this is not the case. They have no desire to escape from their countries. They see their predicament as the consequences of the actions of other states forcing their displacement or migration and not the actions of their own leaders. Their first and primary response is to do whatever is possible to save their island. In this sense relocation elsewhere is seen as giving up and abandoning their home and culture.

Under the leadership of Anote Tong, the Kiribati government developed the *Migration with Dignity* policy.[43] Tong believed in the importance of planned long-term migration and/or relocation as a way of ensuring and maintaining the self-determining agency of the I-Kiribati people should they need to relocate. The policy was not only intended to convey agency through self-determination, resilience and visibility but also to avoid the pitfalls of the I-Kiribati people becoming climate refugees. Tong contends:

> Relocation, no matter how undesirable, must therefore be the brutal reality of the future of atoll island nations, and part of the solution. Kiribati has advocated that migration with dignity must be part of a climate change adaptation strategy, rather than relocation of its people as climate refugees.[44]

---

42. Jane McAdam, 'Refusing 'refuge' in the Pacific: (de)constructing climate-induced displacement in international law', in *Migration and Climate Change*, edited by Etienne Piguet *et al* (Cambridge, UK: UNESCO Publishing & Cambridge University Press, 2011), 116.

43. Republic of Kiribati, 'Fiji Supports Kiribati On Sea Level Rise: Press Release' (11 February 2014) at <http://www.climate.gov.ki/category/action/relocation/>. Accessed 30 January 2021.

44. Anote Tong, 'Migration is the "brutal reality" of climate change', in Climate Home News (21 June 2016) at <https://www.climatechangenews.com/2016/06/21/anote-tong-migration-is-the-brutal-reality-of-climate-change/>. Accessed 30 January 2021.

Tong's policy was twofold with the individual or household as the focal point.[45] In the first instance Tong envisioned a role for the diaspora in helping to resettle migrants in their new home. Tong's hope was that through long term merits-based migration 'pockets' of I-Kiribati communities would be built up overseas that would help facilitate gradual, transitional migration.[46] This would also be a way for I-Kiribati traditions and culture to be kept alive and also enable the I-Kiribati people to slowly adapt to the new host culture and way of life. The second part of the policy was aimed at improving the educational and vocational qualifications that could be obtained in Kiribati to meet the requirements of the countries that residents may migrate to. For Tong the policy was a practical way of affirming agency and avoiding the pitfalls of being identified as refugees that would lead to the loss of voice and rights. This would mean that the future of the I-Kiribati community would be determined by them rather than by others. This vision can only be realised through the support of receiving countries. The limitation of the policy however, is that it only helps pave the way for those who are ready, willing and have resources to migrate, but it does not reach everyone, especially those with very limited literacy skills or those with largely subsistence livelihoods.[47] A further critique of the policy relates to whether or not such a policy will result in long-term positive outcomes in both sending and receiving countries.

Tuvalu holds a contrary view. The former Prime Minister of Tuvalu, Enele Sapoaga is skeptical of the term *Migration with Dignity* stating that there can never be any dignity or human agency in a forced displacement.[48] Where would Tuvaluans go and what would happen to their sovereignty? A focus solely on climate induced migration can easily mask the issues of injustice that have caused the forced migration or relocation in the first place as well as shift the responsibility and focus of the underlying issue at hand. Sapoaga argues that migration implies a choice made by people to migrate; he insists that this

---

45. Karen E McNamara, 'Cross-border migration with dignity in Kiribati', in *Forced Migration Review,* 49 (2015): 62.
46. Jane McAdam, 'Refusing 'refuge' in the Pacific: (de)constructing climate-induced displacement in international law'.
47. McNamara, 'Cross-border migration with dignity in Kiribati', 62.
48. Personal Conversation, Prime Minister of Tuvalu (23 October 2018) (Funafuti, Tuvalu: Office of the Prime Minister).

is not the case for Tuvaluans. Tuvaluans do not choose to move; they will be forced to move. Sapoaga asserts that forced displacement is the correct term for describing the plight of Tuvaluans and is adamant that a distinction between climate induced migration and climate induced displacement is clearly made.[49] Sapoaga also argues that a focus on planned migration undermines the human agency and dignity of climate displaced migrants as well as necessary urgent action on climate change. Sapoaga is of the determined view that there can be no Plan B for Tuvalu. In his view having a Plan B would only serve to the let the culpable emitting countries off the hook. The workshop on climate induced migration *Toku Fenua, Toku Tofi* (My Island, My Birthright) organised by the Tuvalu Association of Non Governmental Organisations (TANGO) held in Tuvalu in June 2018 produced an outcome statement reiterating a long held view that migration will only be considered by Tuvalu as an option of last resort—and only after all other options have been exhausted.[50]

Although Maina Talia supports the position of his government, he believes that this should not stop or limit Tuvuluans from exploring other options. He favours a proactive approach. He asserts that Tuvalu should also explore and have Plan B, C, and D which he outlined in his paper to the Green Party in the German parliament in 2019.[51] Talia argues that exploring other options does not signify abandonment of Tuvalu but rather to be prepared well ahead of time should the unthinkable happen. Talia's paper emphasises the critical importance of agency stressing the importance of consultation with and the participation of the communities involved. He contends:

> It is not surprising for small Pacific Island states like Tuvalu that those who decide our fate have the power and the resources: they are making decisions while they live in a

49. Personal Conversation, Prime Minister of Tuvalu (23 October 2018).
50. *Climate Change Induced Displacement Workshop (Toku Fenua, Toku Tofi: My Island, My Birthright)* (Funafuti, Tuvalu: Tuvalu Asscoiation of Non-Governmental Organisations [TANGO], 2018).
51. Maina Talia, *Migration is a Definite "No" But Rather a Matter of Choice: Voice from the Margin* (Berlin, Germany: Green Party, German Parliament, 2019). Talia outlines four options for exploration, including: Plan B: explore the purchase of land in neighbouring countries, Plan C: explore the possibility of a Climate Passport using the principles and framework of Nansen Climate Passport Scheme and Plan D: explore the concept of Nation *Ex-Situ*.

state of relative safety and with the benefit of what is being called climate privilege. Now, they must surely listen to the voices of those who live in communities that have the very least resources to respond to the negative impacts of climate change and rising sea levels.[52]

A point of contention concerns when climate change displaced migration will become a policy issue to be discussed with possible destinations which are greenhouse emitting countries[53] Although it is recognised that dialogues such as this should happen, this has not always been the case. New Zealand has by far been the exception resolving in November 2017 to consider creating a climate change refugee status for people displaced by rising seas. How this policy is developed and will be implemented remains to be seen.[54] Interestingly, in February 2019, former Prime Minister of Australia, Kevin Rudd in his post Australia Day reflection essay on the challengers facing Australia and Australian identity, identifies climate displacement as a major challenge facing the region. Rudd proposes that:

> Australia should consider developing a proposal to these three states to enter into formal constitutional condominium with them, as we currently have with Norfolk Island. This would require constitutional changes in all four countries. If our neighbours requested this, and their peoples agreed, Australia would become responsible for their territorial seas, their vast Exclusive Economic Zones, including the preservation of their

---

52. Talia, *Migration is a definite "No" But Rather a Matter of Choice: Voice from the Margin*, 8.
53. See John R Campbell and Richard Bedford, 'Migration and Climate Change in Oceania' in *People on the move in a changing climate: the regional impact of environmental climate change on migration*, edited by Etienne Piguet and Frank Laczko (New York, NY, and London, UK: Springer, 2014), 177–204, for an overview of the factors affecting climate migration in the Pacific.
54. For the New Zealand government's developing policy on climate change related migration and displacement, see: New Zealand Foreign Affairs and Trade, *Our Relationship with the Pacific* at <https://www.mfat.govt.nz/en/countries-and-regions/pacific/>. Accessed 30 January 2021. See also Winston Peters, *Pacific climate change-related displacement and migration: A New Zealand Action Plan* (Auckland, NZ: New Zealand Foreign Affairs and Trade, 2018) available at <https://www.mfat.govt.nz/assets/Uploads/Redacted-Cabinet-Paper-Pacific-climate-migration-2-May-2018.pdf>. Accessed 30 January 2021.

precious fisheries reserves. Under this arrangement, Australia would also become responsible for the relocation over time of the exposed populations of these countries (totalling less than 75,000 people altogether) to Australia where they would enjoy the full rights of Australian citizens.[55]

This proposal was of course condemned by the former Prime Minister of Tuvalu Enele Sapoaga as 'imperial thinking' and tantamount to a form of 'neo-colonialism'.[56] But as Bridget Lewis demonstrates, although Australia has not adopted an official policy on climate displacement and relocation in the region, the topic has been raised in a policy discussion paper titled *Our Drowning Neighbours* written by Australian Labour Party MPs, Albert Albanese and Bob Sercombe.[57] On the issue of relocation, Albanese and Secombe proposed the formation of an 'international coalition of nations which would receive citizens of Pacific island nations who are forced to relocate due to the effects of climate change, and for Australia to accept its 'fair share' of 'climate change refugees as part of our humanitarian immigration program'.[58] The discussion paper acknowledged the likelihood of large scale displacement in the future and the need for contingency planning and the need for greater cooperation with the UN to develop a comprehensive legal framework on the issue of 'climate refugees'.[59] But the policy adopted by the Labour Party in 2007, although acknowledging the potential for widespread displacement, 'emphasises taking steps to promote in situ adaptation and disaster risk reduction and does not specifically address the issue of Australia's role in regional migration' nor 'recognise the fact that if migra-

---

55. Kevin Rudd, 'The Complacent Country', at <https://kevinrudd.com/2019/02/04/the-complacent-country/>. Accessed 30 January 2021.
56. Anthony Stewart, 'Tuvalu PM slams Kevin Rudd's proposal to offer Australian citizenship for Pacific resources as neo-colonialism', in *ABC News* (17 February 2019) at <https://www.abc.net.au/news/2019-02-18/tuvalu-pm-slams-kevin-rudd-suggestion-as-neo-colonialism/10820176>. Accessed 30 January 2021.
57. Bob Sercombe and Anthony Albanese, *Our drowning Neighbours: Labor's Policy Discussion Paper on Climate Change* (Canberra, ACT: Australian Labor Party, 2006).
58. Sercombe and Albanese, *Our drowning Neighbours,* 11.
59. Bridget Lewis, 'Neighbourliness and Australia's Contribution to Regional Migration Strategies for Climate Displacement in the Pacific', in *QUT Law Review,* 15/2 (2015): 95.

tion is to happen it requires long-term planning and consultation in order to minimise the negative impacts.[60]

Climate induced displacement raises deep probing theological questions for Pacific Island communities. The enduring legacy of Christian mission has manifested in unhelpful theologies and notions of a 'remote powerful all seeing God who judges the world from afar'.[61] Pacific peoples are deeply religious. More than ninety per cent of people in the region identify as Christian or comparatively their lives have been shaped and informed by the Christian faith framed by a mechanistic understanding of a God who yields absolute power and control over everything.[62] In the past several years Pacific theologians have argued for the continuing need to contextualise Christian faith in local soil. However, as Cecile Rubow and Cliff Bird found in their survey of ecological responses to climate change in Oceania, many churches still hold a literalist view of the bible and believe that God is in control of the weather.[63] Thereby, 'emerging eco-theologies face challengers on two separate fronts, namely the depth and grip of embedded biblical theologies in Oceanian churches, and the place of such theologizing within the climate change discourse of disappearing islands'.[64]

There is a critical need to substantiate faith in a God who is present and immersed in the experiences, culture and context of the people. As Vaai has so aptly articulated, 'for the sake of hope, God should be allowed to be a God of relationships who is 'down to earth' and who suffers alongside the suffering of multiple eco-relationships'.[65] Vaai calls for the revival of 'the compassionate and solidarity images of God for the victims, and to critically address and condemn the powers that enhance climate injustices'.[66] Churches have an important to role to play in preparing communities but in order for 'churches to be

---

60. Lewis, 'Neighbourliness and Australia's Contribution to Regional Migration Strategies, 96.
61. Vaai, '"We are therefore we live" Pacific Eco-Relational Spirituality and Changing the Climate Change Story', 11.
62. Vaai, '"We are therefore we live", 11.
63. Cecilie Rubow and Cliff Bird, 'Eco-theological Responses to Climate Change in Oceania', in *WorldViews*, 20 (2016): 159.
64. Rubow and Bird, 'Eco-theological Responses to Climate Change in Oceania', 164.
65. Vaai, "We are therefore we live", 12.
66. Vaai, "We are therefore we live".

dynamic and effective in addressing the new emerging social issues such as the climate crisis, they have to firstly critically address their faith and theological foundations to meet the challenges of such a crisis'.[67] The exploration of and subsequent development of a theology of place in the context of disappearing islands, grounded not only in the Judeo-Christian tradition but also in Pacific Island indigenous epistemologies is critical, to enable, support and facilitate engagement with the climate realities of the here and now.

Ernst Conradie argues that 'the task of a theology of place is to discern the significance of God's presence in this particular location and time. The question is therefore this: Where is God? What is God doing here?'[68] He contends that 'a theology of place is not merely about our places, but, almost literally, about the place of God . . .' which 'begs further reflection on God's presence in our midst as well as on God's transcendence, beyond any restrictions of place and time'.[69] A theology of place affirms that God is in creation or as Mary Duba so powerfully argues in her thesis on theology of place, 'God is here'.[70] Conradie asserts:

> The deepest intuition of a theology of place has to do with this core Christian insight: 'that the triune Cod is not a distant sky divinity or a deist clockmaker who observes things from above but who does not interfere with what is happening on earth. A theology of place is therefore essentially a theology of the presence of God's Spirit—through Wind, Water, Fire and Life.[71]

The strong affirmation God's presence through the incarnation, the cross and resurrection points to 'a God who risks God's own displacement in order to become good place for the displaced, place for those

---

67. Vaai, "'We are therefore we live", 3.

68. Ernst Conradie, 'Towards a Theology of Place in the South African Context: Some Reflections from the Perspective of Ecotheology', in *Religion & Theology,* 16 (2009): 5.

69. Conradie, 'Towards a Theology of Place in the South African Context, 17.

70. Mary Emily Briehl Duba, *God is Here: A Theology of Place and Displacement* (Doctor of Philosophy Thesis) (Chicago, IL: University of Chicago, 2018).

71. Conradie, 'Towards a Theology of Place in the South African Context: Some Reflections from the Perspective of Ecotheology', 17.

who do not know who or where they are.'[72] A theology of place will help to anchor questions of identity, presence and absence that values the context, cultures and indigenous epistemologies. A theology of place values community and interrelatedness that includes the whole of creation, not just the human family. A theology of place will lead to the questioning of power (control of spaces), and injustices. A theology of place will inevitably explore how God (*Deus Migrator*)[73] is on the move together with people on the move.

But the reach of a theology of place is not limited to the context of the Pacific. It will speak also to contexts outside itself, challenging notions of power and privilege but also reminding us of the obligations of Christian hospitality. To a large extent the possible movements of people will depend on the hospitality of other island nations and beyond to welcome, facilitate and support the resettlement of people. How might the church or public theology speak into this public space and inform, shape and change narratives of home, nation and place? How can the international community foster a sense of home—for the new emerging category of stateless, placeless and landless aliens? Clearly, there are a number of complex strands that need to be woven into a theology of place. In my view there needs to be an ongoing dialogue between theological themes of hospitality, home or place, justice, international law, policies on climate displacement migration and human rights.

Climate induced displacement and relocation is a long-term commitment by all parties involved. Most of all it will require creative imagination and courage to reimagine home and place both as people on the move and people who offer hospitality in receiving countries. The key is long term planning that needs to begin now. 'Countries everywhere need to begin planning today for the looming spectre of climate displacement; every government should have in place not only adaptation plans of action, but displacement plans of action as well.'[74] Leckie et al are confident that the 'ingredients required to solve climate displacement are-for the most part—already in place' and that local and national solutions to climate displacement, which are fully consistent with the human rights of those affected, can be found; that

---

72. Duba, *God is Here: A Theology of Place and Displacement*, 279.
73. Peter Phan, 'Deus Migrator - God the Migrant: Migration of theology and theology of migration', in *Theological Studies*, 77/4 (2016): 845, 845–868.
74. Leckie *at al*, *Climate Change and Displacement Reader*, 4.

is, if political will can be generated to do so'.[75] What will be required is 'appropriate long term planning, targeted resource allocations and well-organised popular movements, can together determine what measures are required, where resources can be found, and ultimately, how to achieve them'.[76] Leckie *et al* believe this is an opportune time for 'governments and civil society the world over to build policies, laws and projects simultaneously mitigate climate change and to ensure that adaptation measures to climate change have at their core the resolution of any displacement that occurs'.[77] Critical to the process of relocation is the inclusion and participation of the communities on the move. As Pacific identity and spirituality is intrinsically intertwined to land or place and faith, a theology of place that integrates pastoral and liturgical practice is needed to accompany the process.

---

75. Leckie *at al*, *Climate Change and Displacement Reader*, 3, 5.
76. Leckie *at al*, *Climate Change and Displacement Reader*, 3.
77. Leckie *at al*, *Climate Change and Displacement Reader*, 3.

Agathon: A Journal of Ethics and Value in the Modern World, Vol 8/2021

Science Does Not Make Ethical Prescriptions: But in the Case of Climate Change Maybe it Should?

Claire Williams

Abstract: The natural sciences are teaching us that much human behaviour is unsustainable and will soon lead to the collapse of major life supporting planetary scale systems. Nowhere is this more obvious than in the case of human induced climate change. This paper aims to examine whether law, as the overreaching institution governing our behaviour, ought to in some way reflect scientific knowledge of human impact on Earth. While this may seem a logical proposition on the surface, it brings up complex ethical and practical questions regarding the role of pure scientific inquiry, and in particular the obligation, if any, that science ought to play in shaping law and policy. I examine the moral difficulties in 'reading' from science how we should conduct our communities and ourselves, as well as the inherent problems with deriving an 'ought' from an 'is,' often referred to as 'Hume's gap,' after eighteenth century philosopher David Hume. For science does not make ethical prescriptions. But in the case of climate change maybe it should.

Introduction

> 'The Earth has its music for those who will listen'
> Reginald Vincent Holmes[1]

It is painfully obvious that Western systems of governance have not responded appropriately to climate change or the environmental crisis.[2] Despite a vast network of environmental protection legisla-

---

1. Reginald Vincent Holmes, *Fireside Fancies* (Ann Arbor, MI: Edwards Brothers, 1955), 27.
2. Anna Greer, 'Towards Climate Justice? A Critical Reflection on Legal Subjectivity and Climate Injustice: Warning Signals, Patterned Hierarchies, Directions for Future Law and Policy', in *Journal of Human Rights and the Environment*, 5/ Special Issue (2014): 103.

tion ranging from international law to local policy, environmental destruction is more widespread than ever before.[3] Law still treats the Earth as an assemblage of distinct fragments, or objects, ownership of which is an individual property right guaranteed by the state.[4] I contend that in the face of a rapidly changing planet law ought to be radically restructured so as to reflect scientific understanding of human impact, rather than the prevailing and now outright absurd notion that unlimited growth on a finite planet is a desirable, equitable or even feasible arrangement.

Laws are an essential part of our society's existence; it is questionable whether we could function without defined limits to human behaviour. Law aims to safeguard an orderly society. Illustrations of law are found in the most basic and complex civilisations.[5] Nonetheless this does not mean we should accept the law without demur.[6] Legal development can be closely linked to social development.[7] As cultural views change and evolve, systems of rules to govern human behaviour adapt. Set in the wider context of social history, the advent of laws and societies are closely connected.[8] As new knowledge and understanding of the Earth and human impact on it come to light, law does have the potential to respond appropriately.[9]

As twentieth century legal philosopher HLA Hart points to *In the Concept of Law*, the law is an arrangement of social decrees, and acts as a boundary between what is and is not acceptable conduct; it

---

3. Klaus Bosselmann, 'Losing the Forest for the Trees: Environmental Reductionism in the Law', in *Environmental Laws and Sustainability*, 2/8 (2010): 2424.
4. Fritjof Capra and Ugo Mattei, *The Ecology of Law: Towards a legal system in Tune with Nature and Community* (Oakland, CA: Berrett-Koehler Publishers, 2015).
5. Peter D Burdon, *Earth Jurisprudence: Private Property and the Environment* (London, UK: Routledge, 2014), 5.
6. Margaret Davies, *Asking the Law Question: The Dissolution of Legal Theory* (Sydney: Lawbook Co, 2nd edition, 2002).
7. Lawrence M Friedman, 'Legal Culture and Social Development', in *Law & Society Review*, 4/1 (1969): 29.
8. H Patrick Glenn, *Legal Traditions of the World* (London, UK: Oxford University Press, 2000); Frederick G Kempin, *Legal History: Law and Social Change* (London, UK: Forgotten Books, 2015 (1963)); Stanley N Katz, *The Oxford International Encyclopaedia of Legal History*, volume 6 (London, UK: Oxford University Press, 2009).
9. This statement leaves aside arguments advanced largely in the work of Karl Marx and Friedrich Engels, that law is a tool of the ruling class, and that power often intercedes between law and justice.

creates an inside of suitable behaviour and an outside of intolerable behaviour.[10] Beyond this basic functioning, law may aspire to preserve ideas such as liberty, equality and justice.[11] The term law itself comes from the Latin word *lex*, meaning a system or body of laws, and *jus*, meaning 'justice,' or 'right.' Science, on the other hand, from Latin scientia, translates as 'knowledge.' Science, broadly speaking, represents an organisation of knowledge encompassing universal truths or laws in the form of testable explanations and predictions regarding the physical world.[12] Patterns, denoted by mathematics, reflect the underlying laws of nature, they provide a scientific blueprint; 'science illuminates the Universe.'[13]

The laws of nature exist whether we discover them or not, hence the term universal laws. All other knowledge is human constructed and construed.[14] And while it may be the human-centric gifts of imagination, morality, our capacity for rational thought and our ability to communicate complex ideas through language and gesture that have the potential to change our cultural course, science is how we know what is real,[15] and thus ought to hold a special place in informing our actions, and in particular the laws and policies which govern our behaviour.

Whether we like it or not, the natural sciences are helping us to comprehend how the planet is reacting to the way we live.[16] As a civilisation we continue to compromise with the laws of physics by

---

10. HLA Hart, *The Concept of Law* (Oxford, UK: Oxford University Press, 3rd edition, 1994).
11. Davies, *Asking the Law Question*.
12. Although Aristotle is generally recognised as the forbearer of the scientific method due to his innovative analysis of logical imputation through demonstration, modern scientific discourse can be attributed almost entirely to the work of Galileo Galilei; see Morris Kline, *Mathematics for the nonmathematician* (New York, NY: Courier Dover Publications, 1985), 284.
13. Danielle Teller, 'There's a Good Reason Americans are Horrible at Science' at <http://qz.com/588126/theres-a-good-reason-americans-are-horrible-at-science/>. Accessed 20 July 2016.
14. It is interesting to note that Edward O Wilson has recently put forward a convincing case that advanced alien life forms would likely possess morality, see Edward O Wilson, *The Meaning of Human Existence* (New York, NY: Liveright, 2014).
15. See Richard Dawkins, *The Magic of Reality: How We Know What's Really True* (London, UK: Random House, 2011).
16. Johan Rockström *et al*, 'Planetary Boundaries: Exploring the Safe Operating Space for Humanity', in *Ecology and Society*, 14/2 (2009): 32.

manipulating Earth's self-regulating systems.[17] It is wilful ignorance to think that we can alter natural phenomena such as the carbon or biochemical cycles, which operate on timescales of millions of years, without severe and adverse consequences. Thus, I maintain that law, as 'the most explicit, institutionalized, and complex mode of regulating human conduct,'[18] ought to be drastically overhauled to reflect our extraordinary yet privileged understanding of the planet's complex yet delicate systems and the unprecedented destructive impact current human practices are having on them.

The issues the above proposition raises are many and complex, both in practice and theory. This paper simply looks to lay the first stone in surveying the theoretical aspect of the problem. The argument has three elements. Firstly, that law aims to govern human behaviour. Secondly, that science is teaching us our behaviour is unsustainable. And finally, that law ought to reflect this fact. For the purpose of this paper, I assume that law aims to govern human behaviour.[19] I focus mainly on investigating the place of science in society, the role of pure scientific inquiry, and the ethical implications of what science is teaching us regarding harmful human activity. My attempt to explore whether we can, or should, 'read' from science how we ought to behave, whether law ought to somehow reflect scientific knowledge, is preliminary at best. My hope is to simply open the space for further exploration and conversation on how law might respond to what the United Nations has described as 'the defining issue of our time.'[20]

The Role of Science in Society

> 'We live in a society exquisitely dependent on science and technology and yet have cleverly arranged things so that almost no one understands science and technology. That's a clear recipe for disaster.'
>
> Carl Sagan[21]

---

17. Anders Wijkman and Johan Rockström, *Bankrupting Nature: Denying Our Planetary Boundaries* (Abingdon, UK: Routledge, Revd ed, 2012).
18. Philip P Wiener, *Dictionary of the History of Ideas* (New York, NY: Macmillan, 1980).
19. This aspect of the proposition is complex and warrants a more in-depth standalone discussion than possible here.
20. United Nations, 'Climate Change' at <https://www.un.org/en/sections/issues-depth/climate-change/>. Accessed 23 June 2020.
21. Carl Sagan, 'Interview', in Anne Kalosh, 'Bringing Science Down to Earth', in *Hemispheres* (October 1994): 99.

Anthropologists generally understand human thought to evolve through specific stages, from mythology to scientific reasoning.[22] Human societies tend to start out with a belief in magical laws, often signified by rituals. Mythology in human societies, James Frazer asserts, is based around ideas of a natural law.[23] Here the term 'natural law' is taken to mean occurrences observed in nature, rather than natural law theory.

EB Tylor has interpreted societies need for myths as an attempt to understand observed natural phenomena. Once it is realised these magical laws are without application, natural laws give way to fables around Gods who control nature, thus leading to the development of religion.[24] Eventually it is understood that nature itself follows laws, and a scientific way of thinking is developed. Frazer contends that myths are made obsolete via science as cultures' advance, 'from magic through religion to science'.[25] Although, as Lucien Lévy-Bruhl is quick to point out, 'the primitive mentality is a condition of the human mind, and not a stage in its historical development'.[26]

However, science will never replace myth, religion or our need to believe, indeed, there is now strong evidence that students of science actually unlearn some of their more basics instincts in the pursuit of scientific knowledge.[27] Nonetheless a sense of wonder at how the universe works is at times such a spiritual experience that it may be likened to religious wonder. Scientists themselves often take science to be 'a way of life . . . a perspective . . . [it] is the process that takes us from confusion to understanding in a manner that's precise, predictive and reliable . . . a transformation, for those lucky enough to experience it, that is empowering and emotional'.[28] This sentiment was echoed by Albert Einstein when he said, 'if something is in me

22. EB Tylor, *Anthropology: An Introduction to the Study of Man and Civilization* (London, UK: Macmillan and Co, 1881).
23. Robert Segal, *Myth: A Very Short Introduction* (Oxford, UK: Oxford University Press, 2004).
24. James Frazer, *The Golden Bough* (New York, NY: Macmillan, 1922).
25. Frazer, *The Golden Bough*, 711.
26. Francois-Bernard Mache, *Music, Myth, and Nature, Or, The Dolphins of Arion* (Abingdon, UK: Taylor & Francis, 1992), 8.
27. Andrew Shtulman and Joshua Valcarcel, 'Scientific Knowledge Suppresses But Does Not Supplant Earlier Intuitions', in *Cognition*, 124 (2012): 209.
28. Brian Greene, 'Put a Little Science in Your Life' at <https://www.nytimes.com/2008/06/01/opinion/01greene.html>. Accessed 3 June 2020.

which can be called religious then it is the unbounded admiration for the structure of the world so far as our science can reveal it'.[29]

Science is not an entrenched aspect of our species' nature the way that religion, sex, altruism and aggression are.[30] We are nonetheless naturally inquisitive creatures, full of '[w]onder, curiosity and delight'.[31] Science, though not inherent to the functioning of human communities, has proved to be an intellectual hallmark of mankind. As Nobel Prize-winning physicist Burton Richter points out, 'this urge to understand is part of the human makeup and should not be ignored'.[32] We are all natural-born scientists. We spend our childhood years exploring the world around us, curious to learn about the environment and what our place is within it. The instinct for pure scientific inquiry comes solely from this curiosity, with no thought as to practical aims or implication.

Humankind has investigated the unknown for at least as long as recorded history. Our lives have been enriched by the knowledge science has uncovered. From the Mohists' backing of the study of logic to Aristotle accompanying Alexander the Great on his marches through Asia to early Abbasid Caliphs sponsorship of the Translation Movement, scientific inquiry has been supported and encouraged.

The launch of the Hubble Space Telescope and bodies such as CERN, Fermilab and SLAC, whose research aims to discover the ultimate structure of matter, demonstrate the continued upkeep of scientific enquiry and the scientific community.[33] Particularly since the end of the Second World War, science has held a comparatively hon-

---

29. *Albert Einstein, the Human Side: New Glimpses from His Archives*, edited by Helen Dukas and Banesh Hoffman (Princeton, NJ: Princeton University Press, 1981), 43.
30. Edward O Wilson, *On Human Nature* (Cambridge, MA: Harvard University Press, 2nd edition, 2004).
31. David Wood, 'Responsibility in An Age of Climate Change', in *Big Ideas*, presented by Paul Barclay (13 May 2015) at <https://www.abc.net.au/radionational/programs/bigideas/responsibility-in-an-age-of-climate-change/6461010>. Accessed 5 January 2021.
32. Burton Richter, 'The Role of Science in Our Society', The Unity of Physics Day Joint Symposium of The American Physical Society and American Association of Physics Teachers (19 April 1995), Washington, DC, USA at <http://slac.stanford.edu/cgi-wrap/getdoc/slac-pub-9284.pdf>. Accessed 1 July 2017.
33. Richter, 'The Role of Science in Our Society'.

oured position in civilisation.[34] Until recently, government funding has been generous and science was viewed as a means to improve our quality of life and keep us secure.[35] In *Science—The Endless Frontier*, a report prepared for President Franklin D Roosevelt, Dr Vannevar Bush recommended that:

> Advances in science when put to practical use mean more jobs, higher wages, shorter hours, more abundant crops, more leisure for recreation, for study, for learning how to live without the deadening drudgery which has been the burden of the common man for ages past . . . Science, by itself, provides no panacea for individual, social, and economic ills . . . But without scientific progress no amount of achievement in other directions can insure our health, prosperity, and security . . .[36]

Science permits us to explore, explain, characterise and ultimately attempt to understand the natural world, which we are also a part of. Contemporary Western society has been shaped by science. Science has allowed us to apprehend, manipulate and exploit the natural world for human endeavour. The clothes we wear, the cars we drive, the music we listen to, the movies we watch, the cities we live and work in; all are a result of scientific discovery. Most often, these technologies came about as a by-product or indirect consequence of pure scientific inquiry. That is, knowledge driven solely by human curiosity. The aim of pure science is simply to increase our understanding of how nature works.

How We Know What Is Real

> 'Science, my lad, is made up of mistakes, but they are mistakes which it is useful to make, because they lead little by little to the truth.'
>
> Jules Verne[37]

---

34. Richter, 'The Role of Science in Our Society'.
35. Richter, 'The Role of Science in Our Society'.
36. United States Office of Scientific Research and Development, *Science—The Endless Frontier: Three Centuries of Science in America* (Manchester, NH: Ayer Company Publishers, 1995 (1945)), 10.
37. Jules Verne, *A Journey to the Center of the Earth: Enriched Classics* (New York, NY: Simon and Schuster, 2008), 47.

Science, like all disciplines, is founded on a knowledge building process. However scientific knowledge differs from other forms of knowledge in its objectivity and verifiability. Scientific conclusions are drawn from physical laws, deduced through observations in the field, experiments in laboratories and mathematical modelling.[38] Scientists examine closely evidence that disagrees with preliminary interpretations.[39] In this way the body of science is constantly growing and improving. Of course, scientists make mistakes like everyone else, but the scientific method is designed to locate and correct them.[40]

The 'laws of nature', broadly speaking, consist of stated regularities, based on repeated experiments or observations, which describe or predict a range of natural phenomena, under a stipulated set of conditions, either universally or in a stated proportion of instances.[41] Though the term 'law' has a diverse usage across all fields of natural science, that is physics, Earth sciences, chemistry and biology, in all cases laws are directly or indirectly based on empirical evidence. The laws of nature, also referred to as scientific laws, are often, although not always, formulated in mathematical language due to the accuracy and internal uniformity that mathematics provides.[42] They are discovered rather than invented.[43]

---

38. PH Gleick *et al*, 'Climate Change and the Integrity of Science', in *Science*, 328/5979 (2010): 689.
39. James Hanson, *Storms of My Grandchildren: The Truth about the Coming Climate Catastrophe and Our Last Chance to Save Humanity* (London, UK: Bloomsbury, 2009).
40. Gleick, 'Climate Change and the Integrity of Science', 689–6790.
41. The term is ambiguous and the topic of much philosophical discussion, see for example John W Carroll, 'Laws of Nature', in *The Stanford Encyclopedia of Philosophy* at <https://plato.stanford.edu/entries/laws-of-nature/>. Accessed 2 May 2020.
42. While scientific laws explain what our senses perceive, they are still empirical and, thus, are not mathematical facts; reference to a law often suggests a fact, although facts do not exist scientifically *a priori*. Many laws, however, do reflect scientific symmetries found in nature and appear as mathematical definitions. Much the same as scientific theories, they are usually referred to as facts. Scientific laws reflect knowledge that has been repeatedly verified through experimentation and never falsified. While scientific theories are falsifiable, mathematical theorems, or identities, such as Pythagorean theorem, possess absolute certainty.
43. William F McComas, *The Language of Science Education: An Expanded Glossary of Key Terms and Concepts in Science Teaching and Learning* (Berlin, Germany: Springer, 2013), 58.

The laws of nature, or scientific laws, must be distinguished from scientific theories. While both are characteristically developed through sound experimental and observational evidence, scientific laws provide a descriptive explanation of nature given particular conditions, while scientific theories aim to explain how nature operates and why specific distinguishing features are demonstrated. Scientific theories may include several scientific laws. For example, Newton's universal law of gravity simply states that two masses in the universe attract each other with a force that is directly proportional to the product of their masses and inversely proportional to the square of the distance between them. The equation for universal gravitation takes the form:

$$F = G \, \frac{m_1 m_2}{2^r}$$

where F is the gravitational force acting between two objects, m^1 and m^2 are the masses of the objects, r is the distance between the centres of their masses, and G is the gravitational constant. Einstein's theory of general relativity explains that the force of gravity arises from the curvature of space and time. General relativity generalises Einstein's theory of special relativity and refines Newton's law of universal gravitation, thus providing a unified description of gravity as a geometric property of space and time, or space-time.[44] A theory will never change into a law no matter how much new evidence comes to light. Laws are used to make theories, not the other way around. Non-scientists are often misled by the fact that preeminent scientific theories are still speculative. However, this does not mean they are unfounded. In fact, a scientific theory is the ultimate verification in science. Scientific theories comprise the most reliable, thorough, and complete form of scientific knowledge.

Karl Marx once wrote, 'all science would be superfluous if the outward appearance and the essence of things directly coincided.'[45] That is, if everything appeared as it seems on the surface, there would be no need for science. Science is not simply a method of specialised

---

44. The mathematics of general relativity are found in Einstein's field equations.
45. Karl Marx, *Capital Volume III*, Part VII, Revenues and their Sources, Chapter 48: The Trinity Formula at < https://www.marxists.org/archive/marx/works/1894-c3/ch48.htm>. Accessed 27 August 2016.

thinking; science represents the best source of truth or knowledge that we have at a given point in time.[46] The laws of nature do not make the outcome predictable.[47] However when conclusions have been methodically and intensely investigated and tested over time, when there appears an emergent recurrent theme in results they become 'well established scientific theories' and may be referred to as facts.[48] Examples of such theories include planetary motion, universal gravity, the big bang, vaccination, electromagnetism, thermodynamics, conservation of mass and energy, molecular bonds and evolution. Climate change now comes under this category. The science is settled, and we have known for some time.[49]

Scientific Understanding of Human Impact (Planetary Boundaries)

> 'We are living on this planet as if we had another one to go to.'
> Terri Swearingen[50]

In 2009, in response to growing concern over the future of the Earth and humankind, a group of distinguished scientists from diverse fields came together to propose a methodical framework aimed at outlining a 'safe operating space for humanity.'[51] Based on relevant expertise, quantifications for seven 'planetary boundaries' were developed including: climate change (CO_2 concentration in the atmosphere <350 ppm and/or a maximum change of +1 W m^{-2} in radiative forcing); ocean acidification (mean surface seawater saturation state with respect to aragonite ≥ 80% of pre-industrial levels); stratospheric

46. Karl Popper, *All Life is Problem Solving* (Abingdon, UK: Routledge, 2001).
47. Brian Cox, *The Human Universe* (Rothesay, UK: William Collins, 2014).
48. Robert McCormick Adams *et al*, 'Climate Change and the Integrity of Science', in Science Magazine (2010) at <http://www.pacinst.org/climate/climate_statement. pdf>. Accessed 10 May 2015. Note the above discussion of scientific 'facts' in footnote 41.
49. Andrea Seabrook, 'Gore Takes Global Warming Message to Congress', in *National Public Radio* (21 March 2007) at <https://www.npr.org/templates/story/story. php?storyId=9047642>. Accessed 4 July 2014.
50. Terri Swearingen, 'Activist Mom Wins Goldman Prize', in *San Diego Earth Times* (August 1997) at <http://www.sdearthtimes.com/et0897/et0897s17.html>. Accessed 20 July 2020.
51. Rockström et al, 'Planetary Boundaries', 32.

ozone (<5% reduction in O_3 concentration from pre-industrial level of 290 DU); biogeochemical nitrogen (N) cycle (limit industrial and agricultural fixation of N_2 to 35 Tg N yr^{-1}) and phosphorus (P) cycle (annual P inflow to oceans not to exceed 10 times the natural background weathering of P); global freshwater use (<4000 km^3 yr^{-1} of consumptive use of runoff resources); land system change (<15% of the ice-free land surface under cropland); and the rate at which biological diversity is lost (annual rate of <10 extinctions per million species).

The study concluded that humankind has already well exceeded three of these planetary boundaries; biodiversity loss, changes to the global nitrogen (N) and phosphorus (P) cycles, and climate change.[52] Human induced pressures on the Earth have reached a point where abrupt disruption of continental to planetary scale systems cannot be excluded.[53] Crossing the thresholds of planetary frontiers is very likely to result in catastrophic non-linear environmental transformation.[54]

While these issues do not exist in isolation from each other, perhaps the most damaging disturbance of the planet's systems will be in the form of human-induced climate change. While climate is in constant flux, for most of human history the atmosphere has con-

---

52. Nitrogen (N) is the Earth's most abundant pure element, comprising of approximately 78% of the atmosphere. Elemental phosphorus (P) is not found as a free element on Earth due to its high reactivity, rather it exists as a mineral in two main forms of its oxidised state; white phosphorus and red phosphorus. Nitrogen and phosphorus are essential nutrients to plant growth and biodiversity health. However, the disruption of the nitrogen (N) and phosphorus (P) biochemical cycles from industrial and agricultural practices has resulted in rapid and often damaging changes to marine and freshwater ecosystems. Human activity now converts more atmospheric nitrogen gas (N_2) into reactive forms than all of Earth's natural processes combined. Much of this reactive nitrogen is released into the atmosphere. However, through precipitation (acid rain) nitrogen falls back to Earth polluting the biosphere, particularly aquatic systems. Excess production of nitrogen and phosphorus also leads to an abundance of these elements into the ocean from runoff, increasing acidification, depleting marine life and pushing ecological systems beyond their limits. Excessive nutrients from agricultural runoff and sewage has contributed to 500 diagnosed dead zones from low oxygen levels in the ocean, or approximately 245,000km^2 globally, around the size of the UK.
53. Rockström *et al*, 'Planetary Boundaries', 32.
54. Rockström *et al*, 'Planetary Boundaries', 32.

tained approximately 275 parts per million (ppm) of carbon dioxide (CO_2). In the last 200 years, largely since the Industrial Revolution, this figure has risen sharply. At our present 407 ppm initial climate impacts such as intensified Arctic warming, disintegration of the West Antarctic Ice Sheet and increased sea level rise have compelled scientists to determine that we are now well above safe levels.[55] Scientists caution that if atmospheric carbon dioxide fails to return to 350 ppm this century, a tipping point could be reached resulting in 'irreversible catastrophic effects' such as melting of the Greenland ice sheet, a shutting down of the thermohaline circulation and enormous amounts of methane release from hastened permafrost melt; thus contributing to further warming.[56]

The Arctic, Antarctic Peninsula and the interior of large continents are already showing ample evidence of a warmer climate.[57] Between 1970 and 2007 satellite documentation of the Arctic region shows sea ice has decreased by around 8.6% per decade. Temperatures in the Antarctic Peninsula have risen 2.5°C in the last fifty years. There is enough water contained in the Greenland ice sheet to raise sea level by 7m, West Antarctica over 5m and the East Antarctic by 50m.[58] Greenland and Antarctica together hold ninety-six per cent of the planet's ice; if these areas become ice free, sea level will rise over 64m.[59] Even a small change in sea level can have disastrous consequences for humanity. If the Earth was to lose even eight per cent of ice cover, most of Louisiana, including New Orleans, will be submerged, along with the Florida Peninsula and low-lying Pacific Island countries such as the Maldives and thousands of Indonesian Islands.[60] Cities such as New York, London and Shanghai would be gravely endangered.[61]

---

55. James Hansen *et al*, in *The Open Atmospheric Science Journal*, 2 (2008): 217.
56. Ibid 217.
57. Lonnie G Thompson, 'Climate Change: The Evidence and Our Options', in *The Behaviour Analyst*, 33/2 (2010): 153.
58. Thompson, 'Climate Change: The Evidence and Our Options', 153–170.
59. John A Church *et al*, 'Sea Level Change', in *Climate Change 2013: The Physical Science Basis. Contribution of Working Group I to the Fifth Assessment Report of the Intergovernmental Panel on Climate Change* (Cambridge, UK: Cambridge University Press, 2013), 1137.
60. Thompson, 'Climate Change: The Evidence and Our Options', 153–170.
61. Thompson, 'Climate Change: The Evidence and Our Options', 153–170.

Global warming will not only raise global temperatures resulting in amplified drought, heatwaves and bushfires; increased water vapour due to a warmer atmosphere will result in other extremes of the hydrological cycle such as heavy rain, intense storms and flood.[62] Sea level rise causes seawater intrusion into the aquifer systems and rivers, which will cause fertile farmland to become saline and ruin fresh drinking water supplies. Changes in global weather structures will have dramatic impacts on natural ecosystems and cause major problems with local food supplies and habitats.

A warming planet is a changing planet. Humanity must now consider its options carefully. In recent years there has been much focus on adaptation as well as mitigation. But we can only adapt to a certain extent, and adaptation is only possible for those of us who have the means to do so. Lonnie Thompson summarises:

> If we don't carefully consider what our options may be, and we haven't really embraced dealing with some of these issues, then maybe we deserve what we get . . . [w]hether we like it or not, the human race is conducting an experiment. We are changing the composition of our atmosphere, and there will be consequences for that . . . [w]e're deciding to just let the climate system do its thing. And it will. I have no doubt that the climate system will take care of the problem. But I don't think we're going to like the way it does it.[63]

As a society we are incredibly privileged to have a large volume of scientific research on the Earth and human impact. Organisations and publications such as the Millennium Ecosystem Assessment Report (MEA), the Intergovernmental Panel on Climate Change (IPCC), the United Nations Environment Program (UNEP) and the Food and Agriculture Organisation (FAO) all contribute relevant, up to date and detailed assessments of different aspects of the Earth's environment. These international assessment bodies were in fact formed for the purpose of briefing lawmakers. In addition, there are many more government institutions, universities and independent scien-

---

62. Hanson, *Storms of my Grandchildren*.
63. Earle Holland, 'Climate Change: Clear and Present Danger' at <http://thinkprogress.org/climate/2011/03/22/207657/lonnie-thompson-global-warming-poses-clear-and-present-danger/>. Accessed 24 November 2016.

tific organisations as well as individuals who continue to add primary research to the growing scientific understanding of planet Earth. This ever-expanding body of science should inform policymakers, economists, and educators, as well as the broader human community in making decisions in relation to the needs of existing generations, future generations and the Earth itself.[64] It should also inform law.

Hume's Gap

> 'What the climate needs to avoid collapse is a contraction in humanity's use of resources; what our economic model demands to avoid collapse is unfettered expansion. Only one of these sets of rules can be changed, and it's not the laws of nature.'
>
> Naomi Klein[65]

In light of what science is teaching us regarding harmful human conduct, law, as an institution that aims to regulate our behaviour, ought to respond to this understanding of the Earth, its planetary boundaries and ecological limits. This is not to say that all laws should somehow directly reflect those found in nature. Simply that in light of a rapidly changing planet, law should provide a framework within which humanity can safely and sustainably operate. By the measured use of human reason we have been able to determine that there are laws that regulate nature. By conscientiously attempting to will in accord this reason within us, it is possible for the law to fulfil the demands of morality.[66]

Science, however, is concerned with the way the world is, not the way the world should be or what is desirable. Science proper strives to be value free, independent of context, interpretation and viewpoint.[67] Science does not make ethical prescriptions. This influential doc-

---

64. Stephen J Turner, *A Global Environmental Right* (Abingdon, UK: Routledge, 2013).

65. Naomi Klein, *This Changes Everything: Capitalism vs. The Climate* (New York, NY: Simon and Schuster, 2014), 21.

66. Garrath Williams, 'Nietzsche's Response to Kant's Morality', in *The Philosophical Forum*, XXX/3 (1999): 201.

67. Nancy Cartwright, 'Popper', in *In Our Time*, BBC Radio Podcast presented by Melvyn Bragg (8 February 2007) at <https://www.bbc.co.uk/sounds/play/b00773y4>. Accessed 5 January 2021.

trine, pioneered by the Vienna Circle and Karl Popper, is still relevant today. Science can inform us. But it cannot save us. For the 'problems that face us are behavioural, they are political, they are the way that we organise ourselves, the way we choose to consume, rather than our technical prowess'.[68] A frustrated Stephen Fry may answer 'no science absolutely is the way to save us as all the other things you mentioned are just rubbish'.[69] It is true science provides an empirical standard of understanding. Regardless of time or place we can have some agreement on science; the same cannot be said for politics or religion. Instead, they tend to divide us.

In *Storms of My Grandchildren* James Hanson concludes that 'the science demands a simple rule: coal must be prohibited'.[70] But as Dale Jamieson points out, 'Hanson often writes as if values can simply be read from science . . . he often seems astonished that political leaders do not obey science's commands'.[71] But it is not possible to 'read' from pure science how we ought to react to multifaceted problems such as climate change.[72] In the end science cannot tell us what to do. As Jamieson puts it, 'science can show us the consequences of our actions, but ultimately we must decide, guided by our values, what to try to bring about . . . Climate science, in our present social context, provokes fundamental questions about how we ought to live and organise our societies that it is powerless to answer'.[73]

Nonetheless, I contend that in light of what science is teaching us regarding harmful human behaviour, law, which aims to regulate our activities, ought, in some way, to respond to such knowledge. That law ought to reflect science, as science is how we know what is real.[74] This proposition, while seemingly logical on its surface, raises complex questions regarding the interface between knowledge, reason and morality. Most noticeably, the proposition bares an

---

68. Stephen Fry, 'Can Science Save Us?', in *The Infinite Monkey Cage*, BBC Radio Podcast presented by Brian Cox and Robert Ince (29 July 2014) at <https://www.bbc.co.uk/sounds/play/b04bn0gp>. Accessed 5 January 2021.
69. Fry, 'Can Science Save Us?'.
70. Hanson, *Storms of my Grandchildren*, 174.
71. Dale Jamieson, *Reason in a Dark Time: Why the Struggle Against Climate Change Failed - and What It Means for Our Future* (New York, NY: Oxford University Press, 2014), 76.
72. Jamieson, *Reason in a Dark Time*.
73. Jamieson, *Reason in a Dark Time*, 76.
74. Dawkins, *The Magic of Reality*.

obvious resemblance to the classic 'is-ought' problem, or what is often referred to as 'Hume's gap'.[75]

In his *Treatise of Human Nature*, published in 1739, David Hume points to the difficulties in moving coherently from factual declarations regarding what 'is' to normative or prescriptive assertions regarding what 'ought' to be with little ontological consideration.[76] That is to say, can an 'ought' be derived from an 'is'. Hume points to the fact that reason does not give us any insight into politics or perceptions of justice. He states:

> In every system of morality, which I have hitherto met with, I have always remark'd, that the author proceeds for some time in the ordinary way of reasoning, and establishes the being of a God, or makes observations concerning human affairs; when of a sudden I am supriz'd to find, that instead of the usual copulations of propositions, is, and is not, I meet with no proposition that is not connected with an ought, or an ought not. This change is imperceptible; but is, however, of the least consequence. For as this ought, or ought not, expresses some new relation or affirmation, 'tis necessary that it should be observ'd and explain'd; and at the same time that a reason should be given, for what seems altogether inconceivable, how this new relation can be a deduction from others, which are entirely different from it. But as authors do not commonly use this precaution, I shall presume to recommend it to the readers; and am persuaded, that this small attention wou'd subvert all the vulgar systems of morality, and let us see, that the distinction of vice and virtue is not founded merely on the relations of objects, nor is perceiv'd by reason.[77]

Taking science to be synonymous with knowledge, it may appear self-evident from the facts that science ought to dictate our response to

---

75. David Hume, *A Treatise on Human Nature* (Oxford, UK: Clarendon Press, 2014 [1739]).

76. Morris B Hoffman, 'Evolutionary Jurisprudence: The End of the Naturalistic Fallacy and the Beginning of Natural Reform?', in *Law and Neuroscience: Current Legal Issues*, 13 (2011): 484.

77. Hume, *A Treatise of Human Nature*, Book III: Of Morals, Part I: Of Virtue and Vice in General, Section I: Moral Distinctions Not Deriv'd From Reason, 521.

the environmental crisis on our doorstep.[78] For if aliens were to arrive from out of space and look at the mess we find ourselves in they would consider us 'monumentally stupid because there is enough energy on Earth that arrives in one hour to serve the economy for a whole year. All we have to do is collect it'.[79] Knowledge of nature must always be a good thing.[80] But what we ought to do with that knowledge is a somewhat more complex question, for it inescapably raises issues around what we truly can know of ourselves, particularly our ability to think and act rationally, as well as our capacity to know what is morally the right or wrong course of action.

Noncognitivism, a school of thought largely evolved from Hume's above statement, denies that 'moral judgments are capable of being objectively true, because they describe some feature of the world'.[81] Noncognitivism suggests that moral knowledge is impossible; that we cannot derive 'ought' from 'is'.[82] Here, JW Harris contends, the concept of objective moral knowledge remains disconcerting for two reasons. First, it generally does not meet the same standards as scientific knowledge; there is no mathematical formulation. Any form of moral truth or knowledge simply cannot be demonstrated to be true in the way that much scientific knowledge can. There is something 'queer' as JL Mackie asserts, in supposing that there could be such a thing as 'moral facts'.[83] Morality is not part of the fabric of the universe.[84] The natural sciences, 'with their procedures of observation and repeatable experiments, can alone tell us what objective facts are 'out there'.[85]

---

78. Brian Cox, 'Can Science Save Us?', in *The Infinite Monkey Cage*, BBC Radio Podcast presented by Brian Cox and Robert Ince (29 July 2014) at <https://www.bbc.co.uk/sounds/play/b04bn0gp>. Accessed 5 January 2021.

79. Eric Idle, Can Science Save Us?' in *The Infinite Monkey Cage*, BBC Radio Podcast presented by Brian Cox and Robert Ince (29 July 2014) at <https://www.bbc.co.uk/sounds/play/b04bn0gp>. Accessed 5 January 2021.

80. For a discussion on this see Brian Cox *et al*, 'Can Science Save Us?', in *The Infinite Monkey Cage*, BBC Radio Podcast presented by Brian Cox and Robert Ince (29 July 2014) at <https://www.bbc.co.uk/sounds/play/b04bn0gp>. Accessed 5 January 2021.

81. Richard T Garner and Bernard Rosen, *Moral Philosophy: A Systematic Introduction to Normative Ethics and Meta-ethics* (New York, NY: Macmillan, 1967), 219–220.

82. JW Harris, *Legal Philosophies* (London, UK: Butterworths, 1997), 12.

83. Cited in Harris, *Legal Philosophies*, 21.

84. Harris, *Legal Philosophies*, 21.

85. Harris, *Legal Philosophies*, 21.

Second is the argument that human societies vary so widely across time and space that what may be considered ethical for one culture may not be for another. Contrary to popular understanding, law is not something that 'looms large over society'.[86] Laws and legal systems are not found outside the society which bought them into being, as 'built into all laws in all places are assumptions about social life, and choices about how things should be done in a culture'.[87] Contemporary examples include marriage equality, abortion and circumcision. Similarly, while the death penalty is considered unacceptable in many countries, others employ it with a near fanatical gusto.[88] The disparity in human laws across time and space demonstrates how little we understand of ourselves, our nature and what ought to be collective moral benchmarks.[89]

However, as Harris notes, Hume was making a common-sense point only.[90] That is, an assertion regarding the relationship of logical necessity between propositions. Hume's statement simply points out an obvious flaw found in some forms of 'that most revered of philosophic weapons,' deductive syllogism.[91] A syllogism comprises three elements, a major premise, for example 'all men are mortal,' a minor premise, Socrates is a man, and a conclusion, 'Socrates is mortal'.[92] The validity of the form (that is the relevant information) plus the truth of the premises gives a true conclusion and thus a sound argument.[93] Based on these premises or assumption one may deduce or induce a conclusion. Here the conclusion follows with logical necessity from the premise. Taking the first and second premise as true, one cannot deny the conclusion without inconsistency.[94] The conclusion is a rearrangement of the information in the premises.[95] However problems

---

86. Kathy Laster, *Law as Culture* (New South Wales: Federation Press, 1997), 1.

87. Laster, *Law as Culture*, 1.

88. Marcelo Gleiser, 'Laws of Man and Laws of Nature', in *National Public Radio* (26 June 2013) at <https://www.npr.org/sections/13.7/2013/06/26/195534987/laws-of-man-and-laws-of-nature>. Accessed 12 March 2017.

89. Gleiser, 'Laws of Man and Laws of Nature'.

90. Harris, *Legal Philosophies*, 13.

91. Harris, *Legal Philosophies*, 13

92. Harris, *Legal Philosophies*, 13

93. Anthony Grayling, 'Logic', in *In Our Time*, BBC Radio Podcast presented by Melvyn Bragg (21 October 2020) at <https://www.bbc.co.uk/sounds/play/b00vcqcx>. Accessed 5 January 2021.

94. Harris, *Legal Philosophies*, 13.

95. Grayling, 'Logic'.

arise when, to use the example Harris gives, in the following syllogism, 'all animals rear their young, men are animals, therefore men ought to rear their young'.[96] Here the conclusion does not follow from the premises, as it contains a copula not found in the premises, that is, the 'ought'.[97] In the case of an inductive argument one goes beyond the information found in the premises.[98]

Reconsidering my argument in this light, the major premise contains the statement 'law aims to regulate our behaviour' and the minor premise, 'science is teaching us our behaviour is unsustainable'. Even if these propositions are true, it does not follow as a matter of logical necessity that law 'ought' to reflect current scientific understanding of human impact. According to Hume it is not obvious how one can move from the descriptive premises 'law aims to govern human behaviour' and 'science is teaching us our behaviour is unsustainable' to the prescriptive statement 'law ought to reflect science'. Hume called for caution against this type of inference in the absence of any explanation of how the prescriptive ought statement follows from the descriptive is statement.

Hume's gap represents the idea that all knowledge is based on either logic and definitions, as in the examples above, or on observation, as is often the case in science. Ought statements cannot be *known* in either of these ways. Thus, if there is no such thing as moral knowledge, ought statements are rendered impossible, or at the least of dubious validity. While this is the position of noncognitivism and moral scepticism, other schools of thought contend that moral truths do exist in the light of goal directed behaviour. In particular the work of Alasdair MacIntyre demonstrates that because ethical language developed in the West in the context of a belief in a human *telos*, that is an end or goal, our moral language, which includes the terms good and bad, functions to reflect the way in which behaviour enables the achievement of that *telos*. While there may be a valid argument that we should not act on climate change, for example the Earth may be better off if we just let nature take its course, it seems safe to assume that collectively we are united in our wish to survive.[99] We can assume that is our purpose, or goal.

---

96. Harris, *Legal Philosophies*, 13.
97. Harris, *Legal Philosophies*, 13.
98. Grayling, 'Logic'.
99. For a discussion on this see Cox et al, 'Can Science Save Us?'.

The point Hume was making must also be put into context within the broader realm of human reasoning.[100] As is often pointed out, the fact that the sun has always risen in the east does not mean it is logically inevitable that it will continue to do so, 'Only mathematical deductions from definitional axioms meet the test of logical necessity.'[101] Hume's problem of induction essentially states that as deductive logic will not and cannot go beyond the data or information contained in the premises one must look for a justification in doing so. Still, in science we might, and regularly do, assume that the patterns one observes in the past will carry on into the future. Hume's gap simply asks us to defend why.

Popper inherited Hume's problem of induction and concluded there was no solution to it. Popper came to his famous views on falsification from the reasonable belief that the only valid kind of logic was deductive logic. Deductive logic has a striking advantage in its way of reasoning such that if the premises are true the conclusion must also be true. However, the problem with deductive logic is that the information in the conclusion really is already contained in the premise. And often in science one would like to go somewhere new.[102] The same can be said for processes of reasoning outside of pure science, as is the case here. Popper rightly pointed out that there was no process for doing so. While various methods may be able to be justified, when one looks at the justifications for these methods it becomes apparent that they depend on further assumptions about the way the world is.[103] The reasoning becomes circular. Thus, Popper maintained that there was no way to universally justify moving from data contained in factual premises to a generalised hypothesis or conclusion without making unwarranted and ungrounded assumptions.[104]

Nonetheless if syllogistic deductive validation is the only evidence of truth one must be sceptical of much of the natural sciences. In this sense Hume was perhaps naïve about how science works.[105] If

---

100. Harris, *Legal Philosophies*.
101. Harris, *Legal Philosophies*, 13.
102. Cartwright, 'Popper'.
103. Cartwright, 'Popper'.
104. Cartwright, 'Popper'.
105. John Worrall, 'Popper', in *In Our Time*, BBC Radio Podcast presented by Melvyn Bragg (8 February 2007) at <https://www.bbc.co.uk/sounds/play/b00773y4>. Accessed 5 January 2021.

however, as Harris so aptly points out, there can be some agreement on the ought proposition of a major premise, and given a factual proposition as a minor premise, one can go one to deduce an ought conclusion.[106] If we agree that law aims to govern human behaviour, and given the fact that science is teaching us our behaviour is unsustainable, we can deduce the conclusion that law ought to in some way reflect this knowledge.

Much like building scientific theories from scientific laws, we should make use of the best source of knowledge we have at a given time. For 'so too over contested questions of morals or politics, we may engage in a discourse which supposes that there is a truth to be found.'[107] What the 'ought' represents is simply a different type of knowledge to what the 'is' represents. Furthermore, if human reason and justified methods of induction, such as observation of past patterns, can be employed by science to gain knowledge of the natural world, there is no reason that similar justified approaches cannot be applied to discovering morality or moral truth.

The Scientists' Dilemma

'Those who have the privilege to know have the duty to act.'
Albert Einstein[108]

Before concluding it is informative to briefly explore how scientists themselves have dealt with the 'is ought' problem. Perceptions vary as to how large a role science and scientists personally should play in shaping policy around climate change. Dale Jamieson notes that, 'It is true that some scientists are barely disguised elitists, impatient with process and politics, and apparently confident that if they could get their hands on the machinery of governance, then whatever problems we face could be solved before dinner'.[109]

However, for the most part, scientists are reluctant to step outside the realm of pure science. This is the role they have prepared and trained for, and understandably, many do not wish to risk jeopardising

---

106. Harris, *Legal Philosophies*, 13.
107. Harris, *Legal Philosophies*, 24.
108. Cited in Jill Baron, *Rocky Mountain Futures: An Ecological Perspective* (Washington, DC: Island Press, 2002), xxvii.
109. Jamieson, *Reason in a Dark Time*, 71.

their scientific integrity. This space of pure scientific discovery is important, and should be respected. However, scientists now realise that all science, no matter how theoretical or abstract it may appear, has practical or technical applications, and the potential to transform society and the world. In the coming century, as the Earth continues to change from human induced stress, scientists may find they have an increasing responsibility to help inform governance and the public.

Events of the last century taught us that science has the power to realise vast change, thus scientists have a duty to proceed with care in research and communication. The Manhattan project and development of the hydrogen bomb exposed the naivety of scientists in considering the consequences of their actions. After World War II, scientists truly became aware of their social and moral obligations as well as the potentially powerful role they had to play in shaping civilization.[110] Just as there are always practical implications to scientific endeavour, the unforseen consequences of climate scientists' findings have reached far beyond the realm of pure science. For as Clive Hamilton observes, as climate scientists go about the business of science, running coupled climate models, measuring sea surface temperature (SST) and mean sea level pressure (MSLP), drilling ice cores and measuring ocean acidification and carbon dioxide concentrations, these scientists are 'unwittingly destabilising the political and social order'.[111]

Climate scientists, however, are typically not practicing science with a view to becoming involved with the repercussions their research may have for society or survival of the planet, although the next generation of climate scientists are certainly aware of this element when they choose their discipline. Lonnie Thompson explains:

---

110. In 1955 concerned scientists signed the Russell-Einstein Manifesto, which highlighted the dangers of nuclear warheads, called for world leaders to find peaceful resolves to international conflict and asked of their colleagues, 'shall we put an end to the human race; or shall mankind renounce war?' The Manifesto also created stimulus for the Pugwash Conferences on Science and World Affairs, an international, independent scientific organisation concerned with the social conscience of scientists. See Bertrand Russell and Albert Einstein, *Russell-Einstein Manifesto* (London, UK: 9 July 1955) at <https://www.atomicheritage.org/key-documents/russell-einstein-manifesto>. Accessed 20 July 2020.

111. Clive Hamilton, *Requiem for a Species: Why We Resist the Truth about Climate Change* (New South Wales: Allen & Unwin, 2010), 18.

> Climatologists, like other scientists, tend to be a stolid group. We are not given to theatrical rantings about falling skies. Most of us are far more comfortable in our laboratories or gathering data in the field than we are giving interviews to journalists or speaking before Congressional committees. Why then are climatologists speaking out about the dangers of global warming? The answer is that virtually all of us are now convinced that global warming poses a clear and present danger to civilization (*sic*).[112]

Many physical scientists are now being forced to step outside the realm of impartial researcher. Some, such as Thompson and Hanson, feel a moral obligation to do this. In practicing good science researchers must remain objective, and not put forward any personal judgment. There is therefore a valid argument that scientists risk losing their scientific integrity if their narrative changes from a description of findings to offering opinions on the consequences of those findings.

However, in the case of climate change this view is far too simplistic. Climate scientists are also parents, friends, lovers and members of society with individual moral and political beliefs, religions, cultures and backgrounds.[113] Here there exists a strong moral argument that we need climate scientists to bring their relevant expertise to the table and actively participate in policy debates and educating the public. But this has proved difficult for scientists to do so. The reaction to climate scientists' increasing effort to convey the urgency of human induced climate change has often been met with disbelief, indignation, ridicule, and at times personal attack. Their findings have been misunderstood, misrepresented, trivialised and largely ignored by the media, wider community, politicians, self-interested industry groups and other ideological factions.

In the face of a catastrophic environmental disaster, it is imperative that scientist be allowed to step outside the realm of pure science, should they wish to do so, without risking their integrity as an objective researcher. The motivation for Hanson to speak out came from fearing for his grandchildren's future. He explains in *Storms* that he was scared his grandchildren would one day say, 'Opa understood

---

112. Thompson, 'Climate Change: The Evidence and Our Options', 153.
113. Jamieson, *Reason in a Dark Time*.

what was happening, but he did not make it clear'.[114] In the case of climate change, there exists a strong moral argument that we need climate scientists to bring their relevant expertise to the table and actively participate in policy debates and educating the public.

There is certainly a growing need for qualified individuals to talk about climate change and the state of the Earth. Science characteristically presents findings based on evidence. In order to maintain their integrity, scientists do not offer opinions or suggest what action should be taken based on their conclusions. However, as Clive Hamilton has noted, 'behind the façade of scientific detachment, the climate scientists themselves now evince a mood of barely suppressed panic'.[115] Traditionally not many scientists would be willing to jeopardise their authority and integrity as scientists to speak out and be labelled as advocates. Yet as mean global temperatures and scientists' concerns grow, as the evidence for non-linear climate change mounts and governance systems fail to adequately respond, a number of scientists have felt an increasing moral obligation to step outside their customary role of objective inquiry.

In order for objective research to maintain its integrity, scientific discovery cannot provide any normative prescriptions. Nonetheless, the inferences and practical consequences of pure science are unavoidable, and in the case of climate change carry grave implications for the Earth, humanity and all other living creatures. If climate scientists feel a moral obligation to step outside the role of impartial researcher in order to warn us of the consequences of our actions, they should be given the freedom to do so without retribution or fear of losing their scientific credibility.

Conclusion

> 'My feelings about climate change are a mixture of awe, hope, despair, frustration and anger.'
>
> Mike Raupach[116]

---

114. Hanson, *Storms of my Grandchildren*, xii.
115. Hamilton, *Requiem for a Species*, viii.
116. Melissa Sweet, 'My Feelings About Climate Change are a Mixture of Awe, Hope, Despair, Frustration and Anger', in *Croakey* (5 February 2015) at <https://croakey.org/my-feelings-about-climate-change-are-a-mixture-of-awe-hope-despair-frustration-and-anger/>. Accessed 23 July 2020.

The appreciation we now have of both the Earth and the Universe, from the complexity of ecosystems to the electromagnetic spectrum, has been driven by our natural curiosity through scientific discovery. It is through science that we start to understand our place on the Earth and in the Universe. We now know that we share our home, the only one of its kind, with all other living beings and that the planet's life supporting systems are at times subject to great change. We now understand that current rates of resource depletion are unsustainable and cannot continue indefinitely without natural support structures, which have allowed civilisation to develop and prosper, breaking down. We also know that human activities are pushing the Earth's systems past their natural limits, and that if we do not drastically alter our behaviour we risk serious and permanent damage to Earth. Science has allowed us to recognise our own adverse impact on the planet.

For the past two centuries, science has primarily been utilised by prosperous individuals and governments as an engine for economic expansion and military power. The world has been transformed radically, both politically and economically and well as socially and environmentally.[117] We cannot continue on current 'business as usual' projections without risking serious collapse of natural life supporting systems. Science helped to create the perilous situation we now find ourselves in, and so arguably has a predominant responsibility in helping communities evolve from an infatuation with unlimited growth and consumption to a sustainable ecological and economic paradigm.

Science cannot and will not ever be able to give definitive answers on complex questions such as how sensitive the Earth's feedback mechanisms are to increased greenhouse gas concentrations. There will never be a finite moment where all science is perfected. It is the nature of science that there will always be uncertainties, in this way science is continuously perusing new knowledge. We ourselves are an evolving species. Knowledge of the natural world and of ourselves grows as we do. We are only just learning how to deal with problems that arose primarily in the twentieth century, that is, a massive boom in population and consumption. While aliens may think us 'monumentally stupid' for relying on buried sunshine rather than col-

---

117. Richter, 'The Role of Science in Our Society.'

lecting it in real time, it is important to note that in many ways we have needed to rely on buried sunshine in order to get us to the point where we have the knowledge and technology to even ask whether science can, or ought to, save us.[118]

Science does not make ethical prescriptions, nor should it. It is not the role of the natural sciences to convey how, as a society, we ought to behave. Science can however provide information necessary to achieving a sustainable civilisation. But ultimately, we must decide what to do with that knowledge. Science can, and should, inform us. As an institution that aims to govern human activity, law and governance systems in particular ought to act in accordance with our scientific understanding of human impact on Earth. How we set up our social and political institutions, how we conduct ourselves, how we behave both individually and as a community, ought to reflect scientific understanding of human impact, as science is how we know what is real.

I suggest that in light of what science is teaching us regarding harmful human behaviour, law ought to reflect science. That we should, as has been recently proposed, 'bolster legal boundaries to stay within planetary boundaries'.[119] Armed with scientific knowledge of human impact on Earth, and the presumption that we wish to survive, it seems logical that law, as the overreaching governing measure of our actions, ought to in some way reflect this knowledge. That is despite the problematic notion of moral truth and our questionable ability to act rationally.

Climate change is merely a symptom of what are ultimately grave underlying social problems.[120] The power imbalance among human beings, as well as our continued destruction of the natural world, must ultimately be redirected to a more sustainable and equitable paradigm. The law, as a social construct, is simply one tool that can be employed to modify cultural practices. By looking to science for guidance, by utilising specific scientific concepts regarding knowl-

---

118. Tony Ryan and Lucy Green, 'Can Science Save Us?', in *The Infinite Monkey Cage*, BBC Radio Podcast presented by Brian Cox and Robert Ince (29 July 2014) at <https://www.bbc.co.uk/sounds/play/b04bn0gp>. Accessed 5 January 2021.

119. Guillaume Chapron *et al*, 'Bolster Legal Boundaries to Stay Within Planetary Boundaries', in *Nature Ecology & Evolution*, 1/3 (2017): 1, 1–5.

120. Cormac Cullinan, *Wild Law: A Manifesto for Earth Justice* (Devon, UK: Green Books, 2003).

edge of the Earth and human impact as a frame of reference, the law may become an essential component in bringing about change, rather than a mechanism that enables destruction of the planet and ultimately ourselves.[121]

---

121. Capra and Mattei, *The Ecology of Law*.

Agathon: A Journal of Ethics and Value in the Modern World, Vol 8/2021

A Four-Step Process for Formulating and Evaluating Legal Commitments Under the Paris Agreement*

*Donald A Brown, Hugh Breakey, Peter Burdon,
Brendan Mackey, Prue Taylor*

Abstract: The Paris Agreement requires each party to prepare, communicate and maintain successive Nationally Determined Contributions (NDCs) that it intends to pursue through domestic mitigation measures. NDCs seek a reduction of greenhouse gas (GHG) emissions so as to hold the increase in the global average temperature to well below 2°C above pre-industrial levels, and to pursue efforts to limit the increase to 1.5°C above pre-industrial levels. In accepting the Paris Agreement, nations have accepted both legally binding non-discretionary duties *and* normative obligations about which nations may exercise some discretion and, in doing so, make explicit the moral (or ethical) responsibilities implicit in those obligations. The four steps that all nations should therefore expressly consider in formulating an NDC are: (1) select a global warming limit to be achieved by the GHG emissions reduction target; (2) identify a global carbon budget consistent with achieving the global warming limit at an acceptable level of probability; (3) determine the national fair share of the global carbon budget, based upon equity and common but differentiated responsibilities and respective capabilities; and (4) Specify the annual rate of its national GHG emissions reductions on the pathway to net zero emissions. These steps serve as a guide for nations when developing and communicating their NDCs. The process also provides guidance for other stakeholders (for example, non-governmental organisations, states and others in the inter-

---

* This article was originally published as Donald A Brown, Hugh Breakey, Peter Burdon, Brendan Mackey, Prue Taylor, 'A Four-Step Process for Formulating and Evaluating Legal Commitments Under the Paris Agreement', in *Carbon and Climate Law Review*, 12/2 (2018). Thanks to the Editors for permission to re-publish this revised version.

national community) through formal processes such the Global Stocktake, who seek to evaluate the level of ambition and fairness entailed by NDCs. Nations are of course free to formulate their NDCs after consideration of issues that go beyond the global carbon budget and equity considerations discussed in this paper, such as, for instance, the obligations of developed nations to assist developing nations with GHG emissions reduction commitments.

Introduction

At the Conference of the Parties (COP) 21 in Paris in 2015, 195 nations adopted the Paris Agreement.[1] Under Articles 2–4 of the Agreement, each Party is obliged to prepare, communicate and maintain successive Nationally Determined Contributions (NDCs) that it intends to pursue through (inter alia) domestic mitigation measures. The primary aim of NDCs is the reduction of greenhouse gas (GHG) emissions so as to hold the increase in the global average temperature to well below 2°C above pre-industrial levels, and to pursue efforts to limit the increase to 1.5°C above pre-industrial levels—a requirement this paper refers to as the *warming limit goal*.[2] While the scope of NDCs extends beyond domestic mitigation to include adaptation, finance and other Agreement requirements (including the use of international markets),[3] this paper's primary focus is on domestic mitigation.[4]

In the Paris Agreement, nations also agreed that mitigation pledges would meet an *equity requirement*,[5] which requires that the NDCs must reflect both the highest possible ambition and the principles of common but differentiated responsibilities and respective capabilities, in the light of different national circumstances, based on

---

1. Conference of the Parties, United Nations Framework Conventionon Climate Change, *Report of the Conference of the Parties on Its Twenty-First Session*, Held in Paris from 30 November to 13 December 2015, FCCC/CP/2015/L.9/Re.1.
2. *Report of the Conference of the Parties on Its Twenty-First Session*, Article 2–4.
3. *Report of the Conference of the Parties on Its Twenty-First Session*, Article 6.
4. *Report of the Conference of the Parties on Its Twenty-First Session*, Article 4.2.
5. *Report of the Conference of the Parties on Its Twenty-First Session*, Article 4.

equity and in the context of sustainable development and efforts to eradicate poverty.[6]

In accepting the Paris Agreement, nations have accepted both legally binding non-discretionary duties *and* normative obligations about which nations may exercise some discretion.[7] Legally-binding, non-discretionary duties that nations agreed to can be found throughout the Agreement text when the mandatory language 'shall' is used. On the other hand, when an obligation is described by using the auxiliary verb, 'should', the obligation is understood to allow some discretion by a nation.[8] However, when nations validly exercise their discretion in the formulation of NDCs, because they are also required as a non-discretionary obligation to provide 'information necessary for clarity, transparency and understanding' when communicating their NDCs,[9] they arguably have responsibilities to explain the basis for judgements made in response to their discretionary obligations. This paper identifies a four-step process that nations should follow in formulating and reporting on their NDCs under the Paris Agreement. In invoking the prescriptive term 'should', here and throughout, our intention is not to create additional obligations beyond those already present in the Agreement, but rather to make explicit the moral (or ethical) responsibilities implicit by both their legally binding non-discretionary and discretionary obligations.

Our recommendations take no final position on how the Paris Agreement's equity requirements (discussed in Step 3) should be interpreted and applied—a subject that is currently debated in a sub-

---

6. This requirement parallels the Agreement's overall purpose, which is to be implemented to 'reflect equity and the principle of common but differentiated responsibilities and respective capabilities, in the light of different national circumstances': Conference of the Parties, United Nations Framework Conventionon Climate Change, *Report of the Conference of the Parties on Its Twenty-First Session*, Article 2, paragraph 2.

7. Daniel Bodansky, 'The Legal Character of the Paris Agreement', in *Review of European, Comparative & International Environmental Law*, 25/2 (2016): 142.

8. The Paris Agreement also creates non-binding discretionary goals (or obligations) for parties by the use of non-normative verbs such as 'will', 'recognise' or 'may' among others. For a discussion of legal significance, see Conference of the Parties, United Nations Framework Conventionon Climate Change, *Report of the Conference of the Parties on Its Twenty-First Session*.

9. Conference of the Parties, United Nations Framework Conventionon Climate Change, *Report of the Conference of the Parties on Its Twenty-First Session*, Article 4.8.

stantial body of literature.[10] Rather, we describe four principles that are widely recognised as valid considerations for interpreting equity, namely: responsibility, equality, capacity, and the right to sustainable development.

The four steps that all nations should expressly consider in formulating an NDC are:

1. Select a global warming limit to be achieved by the GHG emissions reduction target;
2. Identify a global carbon budget consistent with achieving the global warming limit at an acceptable level of probability;
3. Determine the national fair share of the global carbon budget, based upon equity and common but differentiated responsibilities and respective capabilities; and
4. Specify the annual rate of its national GHG emissions reductions on the pathway to net zero emissions.

In addition to the warming limit and equity requirements that apply to all Parties, the Paris Agreement contains certain considerations that differentiate between the obligations of developed and developing countries. These differentiations are beyond the scope of this paper, but warrant further consideration. For instance, only developed nations must initially produce NDCs that are economy-wide, whereas developing nations are required to expand their commitments over time to achieve economy-wide commitments.[11]

The following four sections describe the steps in detail, and identify the sub-issues that have global significance. The subsequent Table 1 describes the minimum information that nations should include with their NDCs.

---

10. On the interpretation of equity, see the references cited in Part IV. Note also: Dirk Messner et al, 'The Budget Approach: A Framework for a Global Transformation toward a Low-Carbon Economy', in *Journal of Renewable and Sustainable Energy*, 2/3 (2010): 031003; Kelly Levin et al, *Designing and Preparing Intended Nationally Determined Contributions (INDCs)*, (New York: World Resources Institute, 2015). Some authors have presented the results of modelled calculations that reflect one or more equity approaches. See, for example, T Wang and J Watson, 'Scenario Analysis of China's Emissions Pathways in the 21st Century for Low Carbon Transition', in *Energy Policy*, 28 (2012): 3537; Yann Robiou du Pont et al, 'Equitable Mitigation to Achieve the Paris Agreement Goals', in *Nature Climate Change*, 7 (2017): 38.
11. Conference of the Parties, United Nations Framework Conventionon Climate Change, *Report of the Conference of the Parties on Its Twenty-First Session*, Article 4, paragarpah 4.

Step 1. Select a Global Warming Limit to be Achieved by the GHG Emissions Reduction Target

The first step in setting a national GHG emissions reduction is to select a warming limit goal (consistent with the Paris Agreement) that the target seeks to contribute to achieving, and then to justify that selection.[12]

The warming limit goal agreed to by the international community in the Paris Agreement is to hold the increase in the global average temperature to *well below* 2°C above pre-industrial levels and *pursue efforts* to limit the temperature increase to 1.5°C above pre-industrial levels. The latter part of this goal was made in the context of recent scientific evidence that a 2°C warming limit may not prevent dangerous climate change. It was also influenced by significant pressure from Small Island Developing States (SIDS) and the most vulnerable least-developed countries. These groups successfully argued that a warming limit above 1.5°C would leave them exposed to dire climate impacts, including in some cases existential threats to their sovereignty.[13] This first step does not enlarge on or limit the application of the Agreement's warming limit goal, but is a recognition that the Agreement creates a warming limit range of between 1.5°C and 2°C.

If a nation formulates a GHG emissions reduction target on a warming limit greater than 1.5°C, it should explain its justification for doing so when it submits its NDC under the Paris Agreement. The nation should explain why the warming limit selected satisfies the nation's responsibility under the UNFCCC to adopt policies that prevent dangerous climate change, and why the nation is unable to set a target that will contribute to achieving a 1.5°C warming limit goal after pursuing efforts to do so.

---

12. If a nation claims that a 1.5°C warming limit would impose unfair costs on it, that claim should be considered when the nation identifies the equity criteria it applied to allocating its fair share of a global carbon budget in Step 3 (see Part IV), rather than when first selecting a warming limit.
13. For a description of the negotiation dynamics that led to agreement on the 1.5°C warming limit, see Wolfgang Obergassel (né Sterk) et al, *Phoenix from the Ashes—an Analysis of the Paris Agreement to the UNFCCC* (Wuppertal, Germany: Wuppertal Institute for Climate, Environment and Energy, 2016), 15, at <https://wupperinst.org/fa/redaktion/downloads/publications/Paris_Results.pdf> accessed 12 June 2017>. Accessed 4 January 2021.

Step 2. Identify a Global Carbon Budget Consistent with Achieving the Global Warming Limit at an Acceptable Level of Probability

To operationalise a global warming limit, it is necessary to identify a global carbon budget, or the total amount of CO_2-equivalent gases (CO_2-e), that can be emitted into the atmosphere before the accumulated stock of atmospheric GHGs exceeds a level that is likely to cause the warming limit to be exceeded. There are scientifically determined relationships between the accumulated stock of GHG in the atmosphere and the level of global warming and associated changes to Earth's climate system. Projections of changes in the climate system are made using a hierarchy of climate models, including complex Earth System Models. The World Climate Research Programme's Working Group on Coupled Modelling coordinates experiments with more than fifty of these models involving over twenty modelling groups performing simulations through the ongoing Coupled Model Intercomparison Project (CIMP). The outputs from CMIP informs the work of the IPCC and provides the scientifically based estimates of global carbon budgets and the likelihood of them achieving a given level of global warming.[14]

Drawing upon CIMP5, the IPCC's 5th Assessment Report (AR5) provided global carbon budgets and associated probabilities (33%, 50% and 66%) for limiting global warming to 1.5°C, 2°C and 3°C above pre-industrial levels. It is important to note that the CMIP's outputs represent multi-model ensemble experiments that provide a consensus representation of the climate system and a measure of the confidence that can be placed in the results. Therefore, the probabilities are better understood as the proportion of all the model simulations that keep warming below that temperature limit.[15] Nonetheless, in practice, it is reasonable to interpret these probabilities as indicating the likelihood that a carbon budget will enable the desired global warming target to be met.

---

14. KE Taylor *et al*, 'An Overview of CMIP5 and the Experiment Design', in *American Meteorological Society*, 3 (2012): 485.

15. Robert McSweeney and Rosamund Pearce, 'Carbon Countdown', in *Carbon Brief* (19 May 2016) at <https://www.carbonbrief.org/analysis-only-five-years-left-before-one-point-five-c-budget-is-blown>. Accessed 19 May 2016.

The IPCC's AR5 Working Group I Summary for Decision Makers Report[16] included the following carbon budgets (Gt CO_2) and probabilities for the period from 1861–1880 to 2011 necessary to limit global warming to less than 2° C: 5,760 Gt CO_2 for >33%; 4,449 Gt CO_2 for >50%; and 3,670 Gt CO_2 for >66%. If forcings from non-CO_2 GHGs are taken into account, these budgets are further reduced to: 3,300 Gt CO_2 for >33%; 3,010 Gt CO_2 for >50% and 2,900 GtCO_2 for >66. The IPCC further reported that 1,890 GtCO_2 was already emitted by 2011 and recent estimates suggest that around 178.7 Gt CO_2 were emitted from 2012–2016,[17] leaving budgets of 1,231 Gt CO_2 for >33%; 941 Gt O_2 for >50% and 831 Gt CO_2 for >66%. A 'ball park' figure therefore, which is supported by recent literature for a >66% chance to keep global average temperature below 2°C above pre-industrial levels, is about 800 Gt CO_2 from 2017.[18] It follows that with each passing year the global carbon budget (i.e., the "permissible" emissions) shrinks as long as anthropogenic greenhouse gas emissions continue and the atmospheric carbon stock grows. Carbon budgets to limit warming to below 1.5°C are of course even smaller with tighter timelines.[19] Updates and commentaries on carbon budgets are being provided on an ongoing basis by the research community and international organisations,[20] and these need to be monitored if NDC commitments are to be informed by the best available science.[21]

---

16. Intergovernmental Panel on Climate Change (IPCC), 'Summary for Policymakers', in *Climate Change 2013: The Physical Science Basis, Contribution of Working Group I to the Fifth Assessment Report of the Intergovernmental Panel on Climate Change*, edited by TF Stocker (Cambridge, UK: Cambridge University Press, 2013), 27.

17. Global Carbon Project, 'Carbon Budget Update' (14 November 2016) at <http://www.globalcarbonproject.org/carbonbudget/>. Accessed 14 November 2016.

18. Ibid.

19. Joeri Rogelj *et al*, 'Energy System Transformations for Limiting End-of-Century Warming to Below 1.5°C', in *Nature Climate Change*, 5/6 (2015): 519.

20. See for example the United Nations Environment Programme (UNEP), *The Emissions Gap Report 2015, a UNEP Synthesis Report* (New York, NY: UNEP, 2015).

21. For example, a paper published in September 2017 found that if the remaining budget to limit warming to the 1.5C Paris Agreement Goal is calculated based upon GHG emissions released so far, the budget is approximately 200 GtC, Miller et al, 'Emission Budgets and Pathways Consistent with Limiting Warming to 1.5°C', in *Nature Geoscience*, 10/10 (2017): 741.

When implementing Step 2, nations also need to appreciate the limitations of the global carbon budgets generated by the CIMP experiments and reported through the IPCC. Climate models are imperfect representations of Earth's climate system and its complex flows of energy, carbon, water and other chemical substances between the atmosphere, ocean, land surface, geosphere and biosphere.[22] The climate-forcing impact of GHGs can only be understood and estimated by taking into account these multiple interacting factors.[23] Current scientific understanding is limited, and known "tipping points" that cause rapid alterations in Earth system dynamics are not yet captured by these global models.[24] While the CIMP multi-model ensemble approach provides some insight into the likely reliability of climate projections, a comprehensive probabilistic understanding requires a synthesis of uncertainties along the cause–effect chain from emissions to temperatures.[25] Limitations in climate models and the uncertainties associated with key climatic process means that the rate of future change is potentially being underestimated. A cautionary approach therefore would support selection of carbon budgets with a lower rather than higher level of uncertainty, e.g., a budget with >66% probability of limiting warming to well below 2°C, versus a budget with a <50%.

Step 3. Determine the National Fair Share of the Global Carbon Budget, Based Upon Equity and Common but Differentiated Responsibilities and Respective Capabilities

Parties to the UNFCCC agreed to:

> protect the climate system for the benefit of present and future generations of humankind, on the basis of equity

---

22. Taylor *et al*, 'An Overview of CMIP5 and the Experiment Design', 485.
23. Intergovernmental Panel on Climate Change (IPCC), *Climate Change 2013: The Physical Science Basis. Contribution of Working Group I to the Fifth Assessment Report of the Intergovernmental Panel on Climate Change* (Cambridge, UK: Cambridge University Press, 2013), 753.
24. James Hansen *et al*, 'Ice Melt, Sea Level Rise and Superstorms: Evidence from Paleoclimate Data, Climate Modeling, and Modern Observations That 2°C Global Warming Could Be Dangerous', in *Atmospheric Chemistry and Physics*, 16/6 (2016): 3761.
25. Malte Meinshausen *et al*, 'Greenhouse-Gas Emission Targets for Limiting Global Warming to 2°C', in *Nature*, 458 (2009): 7242.

and in accordance with their common but differentiated responsibilities and respective capabilities.[26]

These principles were re-committed to in the Paris Agreement.[27] Therefore, a nation should take a position on what the concepts of 'equity', 'common but differentiated responsibilities' and 'respective capabilities' mean, and apply this understanding when calculating its fair allocation of the global carbon budget determined in Step 2. These concepts are usually discussed in relevant climate literature under the term 'equity'.

The Paris Agreement requires that nations supply information about their NDCs necessary for clarity, transparency and understanding[28] and in light of the Article 2 'Purpose of the Agreement'.[29] These requirements were agreed to by the Parties. Therefore, nations have an obligation to be clear, transparent and understandable when responding to all the Agreement's requirements, including its equity requirements.

Although several authors have examined how equity was applied in general terms by nations in formulating their INDCs and several nations claimed to have taken equity into account in formulating their INDCs, no nation explicitly explained how the equity principle they relied upon quantitatively affected their final INDC.[30] In future, this explanation should be provided to achieve the Paris Agreement's information and transparency goals.

As was the case with discussion of the global warming limit goal, this paper is not intended to enlarge on or limit the discretion afforded nations on the Paris Agreement's equity requirement. Rather, we suggest that a nation should explain its application of the Agreement's equity requirement so that the international community can understand how it was applied. This paper therefore examines the equity

---

26. United Nations Environment Program (UNEP), *Rio Declaration on Environment and Development* (New York, NY: UNEP, 1992), A/CONF.151/26 (Vol I), Article 3.1.
27. Conference of the Parties, United Nations Framework Conventionon Climate Change, *Report of the Conference of the Parties on Its Twenty-First Session*, Article 2.
28. *Report of the Conference of the Parties on Its Twenty-First Session*, Article 4, paragarph 8.
29. *Report of the Conference of the Parties on Its Twenty-First Session*, Article13, paragaph 5.
30. National Climate Justice, 'Lessons Learned' (2014) at <https://nationalclimate justice.org/summary-0f-findings/>. Accessed 1 June 2014.

requirements that should guide a nation's GHG emissions reduction obligations under the Paris Agreement. Additional equity issues are entailed by other obligations of developed nations under the Agreement, including the provision of financial resources to assist developing countries with their mitigation, adaptation and technology transfer needs.[31] The magnitude of developed countries' obligations to finance these needs can be determined by considering some of the same equity issues discussed below for developing NDCs. However, these issues are beyond the scope of this paper.

In AR5, the IPCC explained that equity was a fair burden sharing concept,[32] that covers both distributive justice issues and procedural justice.[33] Furthermore, the IPCC said despite ambiguity about what equity means:

> there is a basic set of shared ethical premises and precedents that apply to the climate problem that can facilitate impartial reasoning that can help put bounds on the plausible interpretations of 'equity' in the burden sharing context. Even in the absence of a formal, globally agreed burden sharing framework, such principles are important in establishing expectations of what may be reasonably required of different actors.[34]

The IPCC went on to say that these equity principles can be understood to comprise four key dimensions: responsibility, capacity, equality and the right to sustainable development.[35] Thus, Step 4 requires each nation to determine its allocation of the global carbon budget after giving due consideration to:

---

31. Conference of the Parties, United Nations Framework Conventionon Climate Change, *Report of the Conference of the Parties on Its Twenty-First Session*, Article 9–11, 13.
32. Intergovernmental Panel on Climate Change (IPCC), *5th Assessment Report, Contribution of Working Group III to the Fifth Assessment Report of the Intergovernmental Panel on Climate Change* (Cambridge, UK: Cambridge University Press, 2014), 317.
33. *Assessment Report, Contribution of Working Group III*, section 4.2.2.
34. *Assessment Report, Contribution of Working Group III*, section 4.2.2.
35. *Assessment Report, Contribution of Working Group III*, section 4.2.2.

> *Responsibility*—usually understood to be historical responsibility for current elevated GHG atmospheric concentrations.
>
> *Equality*—each citizen's equal right to use the atmosphere as a sink for its GHGs.
>
> *Capacity*—the ability of a nation to reduce its GHG emissions.
>
> *Sustainable development*—the right of poor countries to pursue economically sustainable development.

Each of these criteria for determining what fairness requires raises questions about how they should be interpreted and applied to determine national mitigation obligations. A brief description of each of these equity principles follows, along with an overview of some of the questions each principle raises—areas over which individual nations have some discretionary latitude.

Responsibility

According to the IPCC, the principle of responsibility can be understood as follows:

> In the climate context, responsibility is widely taken as a fundamental principle relating responsibility for contributing to climate change (via emissions of GHGs) to the responsibility for solving the problem. The literature extensively discusses it, distinguishing moral responsibility from causal responsibility, and considering the moral significance of knowledge of harmful effects. Common sense ethics (and legal practice) hold persons responsible for harms or risks they knowingly impose or could have reasonably foreseen, and, in certain cases, regardless of whether they could have been foreseen.[36]

The principle that a nation should be responsible for reducing GHG emissions in proportion to its historical responsibility for global GHG emissions can be derived from both sound ethical and legal principles. The principle of retributive justice makes those who cause

---

36. *Assessment Report, Contribution of Working Group III*, 319.

harm responsible for the harms that they have caused. Similarly, in international law the 'polluter pays principle'[37] and the 'no harm principle'[38] make those who cause environmental harms responsible for the harms and damages they cause.

There remain questions regarding *when* historical responsibility for past emissions can be triggered, given that human activities such as farming and deforestation have been affecting atmospheric GHG concentrations and the GHG sink capacity of the Earth for thousands of years. For much of this time humans were completely unaware that their activities were affecting GHG atmospheric concentrations and forcing global climate change. Yet by 1990, the year the first IPCC report was published and when negotiations began on a climate treaty, all nations were on notice that human-induced climate change was a growing threat. Therefore, a convincing case can be made that nations should be responsible for, at the very least, emissions after 1990.[39] In any case, we recommend that any government's NDC that interprets equity through the lens of historic responsibility explain on what basis it determined when historical responsibility should be triggered.

Equality

In regard to equality, the IPCC said:

> Equality means many things, but a common understanding in international law is that each human being has equal moral worth and thus should have equal rights. Some argue this applies to access to common global resources, expressed in the perspective that each person should have an equal right to emit.[40]

---

37. UNEP, *Rio Declaration on Environment and Development*, Article 3.1. Principle 16.
38. General Assembly of the United Nations (UNGA), *United Nations Framework Convention on Climate Change* (Rio de Janeiro, Brazil: United Nations General Assembly, 1992), FCCC/INFORMAL/84, GE.05-62220 (E) 200705, Preamble.
39. H Damon Matthews, 'Quantifying Historical Carbon and Climate Debts among Nations', *Nature Climate Change*, 6/1 (2016): 60.
40. IPCC, *5th Assessment Report, Contribution of Working Group III to the Fifth Assessment Report of the Intergovernmental Panel on Climate Change*, 317. Contribution of Working Group III to the Fifth Assessment Report of the Intergovernmental Panel on Climate Change</style></title><short-title>5th Assessment Report (WG III

Although the IPCC acknowledges several interpretations of equality, a case can be made that in determining what equity requires, nations acknowledge an equal right of all humans to use the atmosphere as a sink for GHGs, a consideration that leads eventually to an entitlement to equal per-capita shares.[41] Based on this interpretation, in applying the "equality" principle, nations would determine their fair share of a remaining global carbon budget by allocating the budget on the basis of the nation's percentage of global population— a procedure that would require extremely rapid reductions by high-emitting nations such as the United States. According to the World Bank, the United States has per capita emissions approximately ten times higher than India and almost a hundred times higher than Somalia.[42]

'Contraction and convergence' (C&C) is a well-known equity framework based on allocating an agreed global carbon budget to nations on a per-capita basis and agreeing on the year when national emissions converge on equal per-capita amounts (that is, converge to a common level).[43] The convergence element of C&C acknowledges that high-emitting countries need some time to achieve the required deep cuts in emissions levels. C&C is not a prescription per se, but rather a way of demonstrating how a global prescription could be

41. Anil Agarwal and Sunita Narain, *Global Warming in an Unequal World: A Case of Environmental Colonialism* (New Delhi, India: Centre for Science and Environment, 1991); Paul Baer, 'Equity, Greenhouse Gas Emissions, and Global Common Resources', in *Climate Change Policy: A Survey*, edited by Stephen H Schneider, Armin Rosencranz, and John O Niles (Washington, DC: Island Press, 2002); Niklas Höhne, Michel den Elzen, and Martin Weiss, 'Common but Differentiated Convergence (CDC): A New Conceptual Approach to Long-Term Climate Policy', in *Climate Policy*, 6/2 (2006): 181; Dale Jamieson, 'Climate Change and Global Environmental Justice', in *Changing the Atmosphere: Expert Knowledge and Global Environmental Governance*, edited by P Edwards and C Miller (Cambridge, MA: MIT Press, 2001), 399–408, 191–199, 278–307, 181–99; Peter Singer, 'One Atmosphere', in *Climate Ethics: Essential Readings*, edited by Stephen M Gardiner et al (Oxford, UK: Oxford University Press, 2010).

42. World Bank, 'Co2 Emissions (Metric Tons Per Capita)', in *World Bank* at <http://data.worldbank.org/indicator/EN.ATM.CO2E.PC/countries/1W?display=dW>. Accessed 1 June 2017.

43. Aubrey Meyer, 'Briefing: Contraction and Convergence', in *Proceedings of the Institution of Civil Engineers - Engineering Sustainability*, 4 (2004): 157." <style face="italic">Proceedings of the Institution of Civil Engineers - Engineering Sustainability</style> 157, no. 4 (2004

negotiated and organised in a way that ultimately levels off on the basis of equal per-capita emissions within a constantly shrinking global cap.[44] Implementing C&C requires two steps. First, countries must agree on a warming limit, an associated carbon budget, and the year when global emissions will contract to net zero. Second, countries need to negotiate a convergence date; this date acts as an equity lever because the sooner the date, the greater the mitigation burden carried by countries with high levels of total and per capita emissions.

Even if one acknowledges a right of all people to equal per-capita shares to use the atmosphere, the principle, like many other equity frameworks, does not settle whether basing a nation's allocation on equal per-capita shares should also acknowledge, in its allocation, unequal per-capita use of the atmosphere for *past* emissions. If poor developing nations were required to calculate their NDC for future GHG emissions solely on the basis of equal per-capita shares, they would forego their right to demand that high-emitting rich countries accept both some historical responsibility for past emissions and acknowledge their greater economic capability to pay for higher domestic emissions reductions. Proponents of C&C respond to these concerns by suggesting that these considerations can be taken into account in negotiations between rich and poor countries over economic responsibility to pay for adaptation, climate damage or the costs of reducing GHG emissions in poor countries, along with allowing the international trade of unused allocations, while using the C&C framework to structure future emissions.[45] In the absence of a negotiated resolution of a convergence date between nations, those nations who seek to satisfy their obligations to reduce their GHG emissions, on the basis of a C&C framework by identifying a date by which all nations would converge on equal per capita emissions, should explain the reasoning for the convergence date selected.

---

44. Aubrey Meyer, *Contraction & Convergence: The Global Solution to Climate Change*, Schumacher Briefings (London, UK: Green Books, 2000).
45. Donald A Brown, 'Ethical Problems with Cost Arguments against Climate Change Policies: The Failure to Recognize Duties to Non-Citizens', in *Ethics and Climate* (21 August 2010) at <https://ethicsandclimate.org/2010/08/21/ethical_problems_may_the_nations_rely_on_excessive_costs_to_it_as_justification_for_non-action_on_cl/>. Accessed 21 August 2010.

Capacity

Under the UNFCCC, nations agreed to reduce their GHG emissions to prevent dangerous climate change on the basis of equity and in accordance with their common but differentiated responsibilities and *respective capabilities* (emphasis added).[46] This language was repeated in the Paris Agreement, with the addition of 'national circumstances'.[47] Therefore, a valid consideration for determining a nation's responsibility to reduce its GHG emissions is the nation's capacity to reduce its GHG emissions. In regard to capacity, the IPCC said:

> Generally, capacity is interpreted to mean that the more one can afford to contribute, the more one should, just as societies tend to distribute the costs of preserving or generating societal public goods, i.e. most societies have progressive income taxation. This view can be applied at the level of countries, or at a lower level, recognising inequalities between individuals.[48]

This principle also raises numerous interpretive questions (similar to the issues considered below that are raised by the principle of the right to sustainable development). These questions include: what levels of poverty or gross domestic product (GDP) per capita should be considered in determining a nation's capacity to reduce GHG; whether all GHG emitting activities are excused from reductions by a nation's poverty level (including, for instance, activities like deforestation, gas flaring and wasteful use of electricity generated by fossil fuels); and whether wealthy people's GHG emitting activities in poor countries with low capacity should be excused from GHG emissions restrictions.[49] Nor must (national or individual) wealth be the only variable considered here. As an equity principle, capacity may refer to two linked—but nevertheless distinct—moral considerations. First, capacity can refer to those who can best shoulder economic burdens because of their higher living standards. Due to their existing wealth,

---

46. UNGA, *United Nations Framework Convention on Climate Change*, Article 3.1.
47. Conference of the Parties, United Nations Framework Conventionon Climate Change, *Report of the Conference of the Parties on Its Twenty-First Session*, Article 2.2.
48. IPCC, *5th Assessment Report, Contribution of Working Group III to the Fifth Assessment Report of the Intergovernmental Panel on Climate Change*, 319.
49. Harald Winkler et al, 'What Factors Influence Mitigative Capacity?', in *Energy Policy*, 35 (2007): 692.

new economic and developmental constraints will not cause these people to forego basic necessities. Second, capacity can refer to a particular agent's special capabilities to respond effectively or efficiently to a particular problem. On this footing, 'capacity' could include factors such as technological and scientific expertise, available natural resources, economic structure, institutional capacities, and established governance arrangements. Of course, often these capabilities correlate with wealth, but it would still be helpful for any nation seeking to justify its GHG emissions reductions obligations on low levels of economic capacity, to explain the basis for the distinctions it makes about its capacity to reduce emissions.

Sustainable development

The equity principle based on the right to sustainable development derives from the 1986 United Nations 'Declaration on the Right to Development'.[50] The UNFCCC also acknowledges a right to promote sustainable development, and 'the legitimate priority needs of developing countries for the achievement of sustained economic growth and the eradication of poverty'.[51] The Paris Agreement refers to strengthening the response to climate change and achieving the purpose of the UNFCCC, 'in the context of sustainable development and efforts to eradicate poverty'.[52]

There are several equity frameworks that are consistent with the 'right to develop' principle. One example is the Greenhouse Development Rights Framework (GDR).[53] Specifically designed to ensure that poor people are not unfairly restricted in a world in which the global economy is constrained by limits on GHG emissions,[54] GDR

50. General Assembly of the United Nations (UNGA), *Resolution on Right to Development* (New York, NY, UNGA, 1986), A/RES/41/128.
51. UNGA, *United Nations Framework Convention on Climate Change* (Rio de Janeiro, Brazil: United Nations General Assembly, 1992), Preamble.
52. Conference of the Parties, United Nations Framework Conventionon Climate Change, *Report of the Conference of the Parties on Its Twenty-First Session*, Art 2.
53. P Baer, T Athanasiou, and S Kartha, 'Greenhouse Development Rights: Towards an Equitable Framework for Global Climate Policy', in *Cambridge Review of International Affairs*, 21/4 (2008): 649.
54. Baer, Athanasiou, and Kartha, 'Greenhouse Development Rights'. See also Lasse Ringius, Asbjørn Torvanger, and Arild Underdal, 'Burden Sharing and Fairness Principles', in *International Environmental Agreements*, 2/1 (2002): 1.

specifies that where the average income is below US$7500 per year, people be given the right to development. Under GDR, individuals whose incomes are below this threshold are not expected to help to pay the costs of climate transition. Those with incomes above the development threshold, by stipulation of GDR, are thought of as having realised their right to development.[55] Because of this, under GDR, wealthier nations should shoulder the responsibility of curbing global GHG emissions, the costs of adaptation from unavoidable climate change, and compensation for climate damage.

As would be the case for any framework that makes a distinction on the basis of economic ability to pursue economic development, the GDR has been challenged on the basis of the arbitrariness of the distinctions it makes for allocating responsibility—such as using the US$7,500 threshold to determine who should pay the costs of reducing GHG emissions. In view of this challenge, we recommend that any nation making a distinction on the basis of economic ability to pursue development, explain the basis for the distinction.

The right of developing countries to pursue sustainable economic development raises other interpretative questions such as: how to determine what level of poverty should justify GHG emissions rates lower than the rate reductions required of the entire world community; what limitations on GHG emissions, if any, should be imposed on developing countries who are legitimately pursuing their right to development; and whether developing nations should be expected to limit GHG emissions generated by some human activities such as unnecessary deforestation and degradation, indiscriminate flaring of gases or a high-emitting transportation system. In other words, if one acknowledges the right of poor nations to pursue sustainable economic development, questions remain about what limits on GHGs emission rates should be placed on their development activities. Hence, any nation seeking to justify its future GHG emissions on the basis of its right to sustainable development, should explain the basis for exempting certain activities from the obligations of all nations to reduce their GHG emission to safe levels due to the need to pursue development.

---

55. Baer, Athanasiou, and Kartha, 'Greenhouse Development Rights', 649.

Concluding Considerations on Crafting National Carbon Budgets on the Basis of Equity

Although some authors have recommended one approach to equity,[56] nations are not required to choose only one equity approach. Multiple equity approaches can be combined together to derive different results. Mattoo and Subramanian provide an example of this pluralistic process.[57] In their technical representation, the overall combination of all weighted equity principles sums to 1, meaning (for example) a nation could convert into quantitative results its strong weighting for capacity (0.4), alongside its respect for equal shares (0.3) and needs (0.2), as well as a limited concern for historical responsibility (0.1).

Combining equity approaches makes sense morally. After all, most liberal democracies employ a combination of principles to distribute economic goods. Market principles grant rights to pursue development, while regulations aim at promoting environmental, social and economic sustainability (together achieving sustainable development). Tax policies follow progressive principles (capacity), while welfare measures positively protect the vulnerable (equality and need). Meanwhile, courts and regulatory bodies determine liability for past breaches of criminal, civil and regulatory codes (historical responsibility). Such pluralism is also suggested by the agreements themselves; both the UNFCCC and the Paris Agreement favorably reference multiple principles, making a combined approach a plausible interpretation of the moral responsibilities invoked.

Summing up, climate change presents the global community with a distributive justice problem. As Brown notes, distributive justice 'puts the burden on those who want to move away from equal shares to demonstrate that the justification for their requested entitlement to non-equal shares is based on morally relevant grounds.'[58] As a result, countries require a morally supportable justification when making claims for differential treatment.

---

56. World Resources Institute, 'Climate Equity' at <http://www.wri.org/our-work/project/climate-equity>. Accessed 11 June 2017.
57. Aaditya Mattoo and Arvind Subramanian, 'Equity in Climate Change: An Analytical Review', in *World Development*, 40/6 (2012): 3583.no. 6 (2012
58. Brown, 'Ethical Problems with Cost Arguments against Climate Change Policies: The Failure to Recognize Duties to Non-Citizens'.

Although reasonable people can disagree on what fairness requires of nations when allocating national shares of a carbon budget, not all claims made by nations about the fairness of their GHG emissions targets are ethically valid or meritorious. Following Amartya Sen, people do not have to reach agreement on what perfect justice requires to obtain agreement on injustice.[59] Thus, there is hope that some progress can be made on the issue of what fairness and equity requires of nations in the stocktake meetings that will take place under the Paris Agreement in the years ahead, provided nations are required to specifically explain how equity considerations affected the calculation of their NDC. On the basis of the equity principle(s) selected, a nation can calculate the portion of the remaining global carbon budget that should be allocated to it. This calculation must be clear and transparent so that all parties can evaluate how the equity principle affected the NDC.

Overall, Step 3 involves the following sub-steps:

1. Nations select the equity principles they will employ in determining their NDC. Combinations of these principles, with different weightings, are possible. For example, a nation might opt for "equal per capita shares", but also take into account its own specific historical responsibility.
2. Nations make decisions about the variables within each principle, and explain their decisions. For example, historical responsibility requires a decision to be made about the starting date for holding nations responsible, while sustainable development requires a decision to be made about the income level below which privileged carbon entitlements will be allocated.
3. Nations employ their selected (combination of) equity principles, with the selected variables inserted, to calculate the numerical estimates of their national share of the global carbon budget.

Step 4. Specify the Annual Rate of the National GHG Emissions Reductions on the Pathway to Net Zero Emissions

Once a national carbon budget has been determined in Step 3, the nation can then calculate the annual rates of reductions that will

---

59. Amartya Sen, *The Idea of Justice* (Cambridge, MA: Harvard University Press, 2011), 141.

define the contraction pathway to net zero emissions. Article 4.1 of the Paris Agreement commits Parties to reach global peaking of GHG emissions as soon as possible, recognising that peaking will take longer for developing country Parties, and to undertake rapid reductions thereafter in accordance with the best available science, so as to achieve a balance between anthropogenic emissions by sources and removals by sinks of GHGs in the second half of this century, that is, net zero emissions.

Clearly, the later the peak and the slower the rate of reduction now, the faster the rate of reduction must be in the future. A nation's emissions reduction target should be understood as the total amount of GHGs (typically expressed in Gt CO_2e) that can be emitted over the full contraction period, not simply as a GHG emissions level at a certain future date.

In determining its national GHG emissions reduction pathway, a nation should be sensitive to the fact that, to give the world a realistic hope of preventing dangerous climate change, deep rapid early emission cuts are necessary. It should also be aware that the longer it waits to reduce emissions, the greater the proportion of the remaining global carbon budget being consumed, and the less likely the probability of meeting global warming limits.[60]

The national emission pathway is a required part of the process for two reasons. First, for the nation itself to move forward on achieving the required mitigation, determining the pathway provides it with a long-range plan of action, keeping in mind that the arc of the mitigation pathway impacts on the overall amount of carbon released (making early action critical). Second, determination and communication of the pathway demonstrates to the international community that the pathway is feasible, and sets down stages of implementation that allow the international community to track the nation's progress. The Agreement provides for progress to be tracked over both short- and long-term periods via mechanisms such as the biennial reporting mechanisms[61] and long-term low GHG emission development strategies.[62] The overall feasibility and staged pursuit of the NDC

---

60. Hansen, 'Ice Melt, Sea Level Rise and Superstorms', 3761.
61. Conference of the Parties, United Nations Framework Conventionon Climate Change, *Report of the Conference of the Parties on Its Twenty-First Session*, Article 13, paragraph 7.
62. *Report of the Conference of the Parties on Its Twenty-First Session*, Article 4.19.

commitment allows the international community to proceed in the expectation that the nation is fulfilling its obligations in good faith. It therefore facilitates collective momentum and reciprocal efforts by other Parties.

That said, the crucial factor remains that the nation keeps within the bounds of its national carbon budget set down in Step 3. Indeed, different domestic political parties might favor different means of achieving the target—a socio-economic question about which reasonable citizens might disagree. Even in the case of differing policies (at least in the details) of Step four's pathway, nations will still have to ensure an overall policy coherence, to guarantee that the attempted imposition of different and conflicting mitigation policies (across electoral cycles) does not stymie the required outcome (*viz*, the successful pursuit of the nation's carbon budget from Step 3).

Conclusion

This paper aims to help governments and stakeholders understand certain key steps that a national government should follow in determining its national GHG reduction targets. The paper identifies information that, if documented and submitted, would allow independent review of the adequacy of a national government's NDC emission reduction targets relevant to the Paris Agreement's warming limit goal and equity requirements. This information is necessary to enable the Agreement's stocktake, transparency and (potentially) compliance mechanisms, to achieve their intended goals.

The four-step process proposed here can be considered as a guide for nations when developing and communicating their NDCs. The process also provides guidance for other stakeholders (e.g. non-governmental organisations, states and others in the international community) through formal processes such the Global Stocktake, who seek to evaluate the level of ambition and fairness entailed by NDCs. Nations are of course free to formulate their NDCs after consideration of issues that go beyond the global carbon budget and equity considerations discussed in this paper, such as, for instance, the obligations of developed nations to assist developing nations with GHG emissions reduction commitments. However, as a minimum standard, achieving the Paris Agreement's warming limit goals on the basis of equity will be facilitated by national governments clearly explaining in their NDCs how they applied the four steps discussed here.

Table 1. Summary of Information Nations Should Submit with their NDC

1.   The global warming limit the NDC is intended to achieve in cooperation with others, and the nation's justification for any warming limit between 1.5°C and 2.0°C, including an explanation of national efforts to achieve a warming as close as possible to 1.5°C.
2.   The carbon budget to achieve the temperature limit for the entire world that the nation's emissions target seeks to implement, including: • the total quantity of carbon in $GtCO_2$ that all the world's nations in aggregate can emit before atmospheric concentrations exceed levels that result in the warming limit being exceeded; • the probability level associated with the carbon budget; • the equilibrium climate sensitivity assumed in calculation of the budget;
3.   The equity formula followed in formulating the GHG target submitted with the NDC, including a description of how the following equity considerations were used in quantitative determination of the national emissions reduction target: A. Historical responsibility: • the year from which historical responsibility was triggered; and, • the reason for triggering historic responsibility at the designated year. B. Equal per capita emissions: • the date at which nations should achieve equal per-capita emissions, and the explanation for the date. C. The right to sustainable development: • the economic criteria that triggered an equitable adjustment to the nation's responsibility on the basis of right to development; and • the justification for this distinction. D. Economic capacity to reduce GHG emissions: • the economic threshold for making an equitable distinction on the basis of economic capacity; and • the justification for this threshold.
4.   The national GHG emissions reduction target in $GtCO_2$, including the rates of reduction and the date when the nation's emissions will need to reach zero.

5.	For developing countries (in addition to the above): <ul><li>GHG emissions reductions that will be achieved by a nation; and</li><li>GHG emissions reductions that will be implemented through financing by developed countries, and the justification for the distinction.</li></ul>
6.	For developed nations (in addition to the above): <ul><li>descriptions and explanation of how the nation responded to its obligations to finance mitigation, adaptation and technology transfer in developing countries.</li></ul>

Agathon: A Journal of Ethics and Value in the Modern World, Vol 8/2021

A Matter of Choice: Property and the Person[*]

Paul Babie

Abstract: This essay argues that the liberal conception of private property is not a solution to anthropogenic climate change but, rather, the source of the problem. The concept of private property facilitates the human activities that cause anthropogenic climate change and the resulting human externalities suffered disproportionately by those in the developing world. The essay concludes that we can no longer wait for government to act; we must take individual, personal, action now if we are to address the challenge of climate change. It is a matter of choice, and the choice is ours.

Introduction

Climate change is a private property problem. Some may react strongly to such a bold claim—after all, private property is seen as part of the solution to the challenge of climate change, using 'commodification'[1] and 'propertisation'[2] of carbon in 'cap-and-trade' schemes.[3] 'Putting a

---

[*]   Earlier versions of this article were published as Paul Babie, 'Climate Change: Government, Private Property, and Individual Action', in *Sustainable Development Law & Policy*, 11/2 (2011): 19–21, 77–78 and Paul Babie, 'Private Property: The Solution or the Source of the Problem?', in *Amsterdam Law Forum*, 2/2 (2010): 1–5. Thanks to the Editors of both journals for permission to re-publish those earlier versions and this revised version.

[1.]   See Margaret Jane Radin, *Contested Commodities: The Trouble with Trade in Sex, Children, Body Parts and Other Things* (Cambridge, MA: Harvard University Press, 1996).

[2.]   See Kevin Gray, 'Property in Thin Air', in *Cambridge Law Journal*, 50 (1991): 252.

[3.]   See, for example, *American Clean Energy and Security Act (Waxman-Markey)*, H.R. 2454, 111th Cong. (2009) at <http://www.govtrack.us/congress/billtext.xpd?bill=h111-2454>. Accessed 5 January 2021.

price on carbon' is of course the adjunct to a cap-and-trade scheme.[4] Still others argue that what we really need is an immediate transition to a low carbon economy.[5]

There is no shortage of warnings. In late 2018, the Intergovernmental Panel on Climate Change reminded the world that climate change is now very close to the point of no return, to causing catastrophic consequences from which we may never recover. Limiting global warming to 1.5°C above pre-industrial levels would avoid many of the catastrophic consequences.[6] These graphs demonstrate visually what must be achieved, now, if we are to meet the 1.5°C target which will avoid climate catastrophe:[7]

Chart 1: Required emission pathways

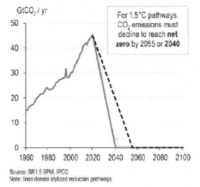

$GtCO_2$ / yr

For 1.5°C pathways, CO_2 emissions must decline to reach **net zero by 2055 or 2040**

Source: SR1.5 SPM, IPCC
Note: lines denote stylized reduction pathways

Chart 2: Required cumulative emissions

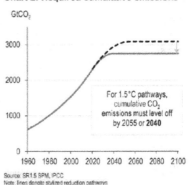

$GtCO_2$

For 1.5°C pathways, cumulative CO_2 emissions must level off by 2055 or **2040**

Source: SR1.5 SPM, IPCC
Note: lines denote stylized reduction pathways

Yet notwithstanding the range of structural economic options available to address the challenge, governments continue to dissemble and avoid taking bold action, of any kind. For many people, the world

4. The World Bank, *Pricing Carbon* (22 June 2020) at <https://www.worldbank.org/en/programs/pricing-carbon>. Accessed 5 January 2021.

5. See Jeremy Rifkin, *The Third Industrial Revolution: How Lateral Power is Transforming Energy, the Economy, and the World* (New York, NY: Palgrave Macmillan, 2013); Jeremy Rifkin, *The Green New Deal: Why the Fossil Fuel Civilization Will Collapse by 2028, and the Bold Economic Plan to Save Life on Earth* (New York, NY: St Martin's, 2019).

6. Intergovernmental Panel on Climate Change (IPCC), *Global Warming of 1.5°C – Summary for Policymakers* (New York, NY: IPCC, 2018) at <http://www.ipcc.ch/report/sr15/>. Accessed 5 January 2021.

7. Intergovernmental Panel on Climate Change (IPCC), *Global Warming of 1.5°C*.

over, this fact causes real alarm.[8] And it ought to, for this governmental failure stands as a depressing indictment of the effects on people of anthropogenic climate change. Over a decade ago, Bjørn Lomborg, the self-proclaimed 'skeptical environmentalist',[9] wrote:

> The risks of unchecked global warming are now widely acknowledged: a rise in sea levels threatening the existence of some low-lying coastal communities; pressure on freshwater resources, making food production more difficult in some countries and possibly becoming a source of societal conflict; changing weather patterns providing favourable conditions for the spread of malaria. To make matters worse, the effects will be felt most in those parts of the world which are home to the poorest people who are least able to protect themselves and who bear the least responsibility for the build-up of greenhouse gases . . . Concern has been great, but humanity has so far done very little that will actually prevent these outcomes. Carbon emissions have kept increasing, despite repeated promises of cuts.[10]

What has changed in the decade since? Nothing. As I wrote in my Editorial to this special issue, the Australian summer of 2019–2020, to draw upon but one example, witnessed 'a terrible trifecta of heatwaves, drought and bushfires, made worse by climate change',[11] palpable evidence of '[a] long-term warming trend from the burning of coal, oil and gas is supercharging extreme weather events, putting

---

8. Andrew Rzepa and Julie Ray, 'World Risk Poll Reveals Global Threat from Climate Change', in *Gallup Blog* (6 October 2020) at <https://news.gallup.com/opinion/gallup/321635/world-risk-poll-reveals-global-threat-climate-change.aspx>. Accessed 5 January 2021.

9. Bjørn Lomborg, *The Skeptical Environmentalist: Measuring the Real State of the World* (Cambridge, UK: Cambridge University Press, 2001).

10. Bjørn Lomborg, 'Introduction', in *Smart Solutions to Climate Change: Comparing Costs and Benefits*, edited by Bjørn Lomborg (Cambridge, UK: Cambridge University Press, 2010), 1–8, 1.

11. Climate Council, *Dangerous Summer: Escalating Bushfire, Heat and Drought Risk* (Sydney: Climate Council, 2019) at <https://www.climatecouncil.org.au/resources/dangerous-summer-escalating-bushfire-heat-drought-risk/>. Accessed 5 January 2021. See also Joëlle Gergis, *Sunburnt Country: The History and Future of Climate Change in Australia* (Melbourne: Melbourne University Press, 2018).

Australian lives, our economy and our environment at risk.'[12] The same trends are found the world over.[13]

Another way of looking at humanity's inaction may simply be the recognition, by governments if not yet by humanity as a whole, that what is necessary is nothing short of wholesale change to the dominant concept of private property. This brief article aims to explain why private property, touted so often as the saviour to the challenge posed by climate change, may in fact be the source of the problem. What we may really need—a new way of choosing—lies hidden in private property, but it is not private property itself. Let me explain.

What Private Property Is

Liberal theory bequeaths us the dominant modern concept of private property.[14] Concerned with the establishment and maintenance of a political and legal order, liberalism, among other things, tries to secure individual freedom in choosing a 'life project'—the values and ends of a preferred way of life[15]—and ongoing choice in how we will furnish ourselves with what we need to pursue that project. In order for life to have meaning, some control over the use of goods and resources is necessary; private property is liberalism's means of

---

12. Climate Council, *Dangerous Summer: Escalating Bushfire, Heat and Drought Risk*.
13. 'Paris-anniversary Climate Pledges Bring Progress But Fall Short', in *The Economist* (13 December 2020) at <https://www.economist.com/international/2020/12/13/paris-anniversary-climate-pledges-bring-progress-but-fall-short>. Accessed 5 January 2021.
14. See Alan Ryan, 'Self-Ownership, Autonomy and Property Rights', in *Property Rights*, edited by Ellen Frankel Paul, Fred D Miller, Jr, and Jeffery Paul (Cambridge, UK: Cambridge University Press, 1994), 241–258; Gerald F Gaus 'Property, Rights, in *Property Rights*, edited by Ellen Frankel Paul, Fred D Miller, Jr, and Jeffery Paul (Cambridge, UK: Cambridge University Press, 1994), 259–286; Joseph William Singer, 'How Property Norms Construct the Externalities of Ownership', in *Property and Community*, edited by Gregory S Alexander and Eduardo M Peñalver (Oxford, UK: Oxford University Press, 2010), 66–70.
15. See Michael J Sandel, 'Introduction', in *Liberalism and its Critics*, edited by Michael J Sandel (New York, NY: New York University Press, 1984), 1; JW Harris, *Legal Philosophies* (Oxford, UK: Oxford University Press, 2nd edition, 2004), 277–300.

ensuring that individuals enjoy choice over goods and resources so as to allow them to fulfil their chosen life.[16]

The liberal conception of private property is, in simple terms, then, a 'bundle' of legal relations (or rights) created, conferred and enforced by the state (law), between people in relation to the control of goods and resources.[17] At a minimum, these rights typically include use, exclusivity, and disposition.[18] One can use one's car (or, with few exceptions, any other tangible or intangible good, resource, or item of social wealth), for example, to the exclusion of all others, and may dispose of it. And the holder may exercise these rights in any way they see fit, to suit personal preferences and desires.[19] Or, we might put this in a way that comports more with the language of liberal theory—rights are the shorthand way of saying that individuals enjoy *choice* about the control and use of goods and resources in accordance with and to give meaning to a chosen life project.

Notice, though, that in my definition, such rights exist only as a product of relationship between individuals. This is significant, for it focuses our attention on the fact that where there is a right—or, as I am putting it, choice—to do something, there is a corresponding duty—or a lack of choice—to refrain from interfering with the inter-

---

16. See Jeremy Waldron, *The Right to Private Property* (Oxford, UK: Oxford University Press, 1988); Stephen R Munzer, *A Theory of Property* (Cambridge, UK: Cambridge University Press, 1990); Margaret Jane Radin, *Reinterpreting Property* (Chicago, IL: University of Chicago Press, 1993); Joseph William Singer, *Introduction to Property* (Alphen aan den Rijn, Netherlands and Philadelphia, PA: Wolters Kluwer, 5th edition, 2016), 2.

17. See Singer, *Introduction to Property*, 2.

18. Radin, *Reinterpreting Property*, 121–123. This builds, of course, upon the groundbreaking work of Anthony M Honoré, 'Ownership' in *Oxford Essays in Jurisprudence*, edited by AG Guest (Oxford, UK: Oxford University Press, 1961), 105–47, who identified eleven 'standard incidents' of ownership.

19. This begins with John Stuart Mill's 'self-regarding act': John Stuart Mill, *On Liberty*, edited by Gertrude Himmelfarb (New York, NY; Penguin, 1974 [1859]). See especially Singer, 'How Property Norms Construct the Externalities of Ownership', 66–70, and Munzer, *A Theory of Property*, 3–9.

est protected by the right.[20] Rights would clearly be meaningless if this were not so. The liberal individual holds choice while all others— the wider community in which one lives, or society—are burdened with a lack of it as concerns that good or resource. C Edwin Baker summarises the idea of rights and relationship as applied to private property this way:

> ... [it] [i]s a claim that other people ought to accede to the will of the owner, which can be a person, a group, or some other entity. A specific property right amounts to the *decisionmaking authority* of the holder of that right.[21]

Private property, then, is not merely about the control and use of goods and resources, but also, significantly, about controlling the lives of others; it is the power we have to choose what others may or may not do with the things said to be 'mine'.[22] Using evocative and graphic language, Roberto Mangabeira Unger puts it this way:

> [t]he right [choice] is a loaded gun that the rightholder [the holder of choice] may shoot at will in his corner of town. Outside that corner the other licensed gunmen may shoot him down. But the give-and-take of communal life and its characteristic concern for the actual effect of any decision upon the other person are incompatible with this view of right ...[23]

Identifying the importance of relationship reveals the fact that private property and non-private property rights overlap; choices made by

---

20. Wesley Newcomb Hohfeld, 'Some Fundamental Legal Conceptions as Applied in Judicial Reasoning', in *Yale Law Journal*, 23 (1913): 16; Wesley Newcomb Hohfeld, 'Some Fundamental Legal Conceptions as Applied in Judicial Reasoning', in *Yale Law Journal*, 26 (1917): 710; Wesley Newcomb Hohfeld, *Fundamental Legal Conceptions as Applied in Judicial Reasoning* (New Haven, CT: Yale University Press, 1919); Wesley Newcomb Hohfeld, *Fundamental Legal Conceptions as Applied in Judicial Reasoning, II*, edited by Walter Wheeler Cook (New Haven, CT: Yale University Press, 1923).
21. C Edwin Baker, 'Property and its Relation to Constitutionally Protected Liberty', in *University of Pennsylvania Law Review*, 134 (1986): 741, 742–743 (emphasis added).
22. Morris R Cohen, 'Property and Sovereignty', in *Cornell Law Quarterly*, XIII (1927): 8, 13.
23. Roberto Mangabeira Unger, *The Critical Legal Studies Movement* (Cambridge, MA: Harvard University Press, 1983), 36.

those with the former have the potential to create negative outcomes—consequences, or what economists call 'externalities'—for those with the latter.[24] At the highest level of generality, Unger's 'gunman' is vested with absolute discretion to '... an absolute claim to a divisible portion of social capital[]' and that '[i]n this zone the rightholder [can] avoid any tangle of claims to mutual responsibility.'[25] The individual revels in '... a zone of unchecked discretionary action that others, whether private citizens or governmental officials, may not invade.'[26]

Every legal system acknowledges this problem and therefore accepts that with rights—choice—come obligations, too, which we owe towards others.[27] And it is the state which, through law, creates private property just as through that same law it regulates or limits what can be done in the conferral of choice. In this way, the state is said to mediate the socially contingent boundary between private property and non-property holders. This is, in fact, the essence of private property—the state confers self-serving rights—choice—upon individuals (including the most unusual of all individuals, the corporation![28]) that come with obligations—the absence of choice—towards others.[29]

Yet there is something much more disturbing lurking just below the surface of what appears to be state control aimed at preventing harmful outcomes like those of climate change. What is really being conferred by private property is what Duncan Kennedy calls the legal ground rules giving permissions to injure others, to cause legalised injury.[30] This is insidious, for '... we don't think of [them] as ground rules at all, by contrast with ground rules of prohibition. This is Wesley Hohfeld's insight: the legal order permits as well as prohibits, in the

---

24. See Singer, 'How Property Norms Construct the Externalities of Ownership', 59.
25. Mangabeira Unger, *The Critical Legal Studies Movement*, 37–38.
26. Unger, *The Critical Legal Studies Movement*, 38.
27. Singer, 'How Property Norms Construct the Externalities of Ownership', 59.
28. See Paul Babie, 'A Tale of Two Creations—Property and the Corporation' in *Comparative Reflections on the Constitutional Models of India and Australia*, edited by Aditya Tomer, Vaishali Arora, Paul Babie, and Lorne Neudorf (New Delhi, India: Bloomsbury, 2020), 2.1-2.16; Paul Babie, 'Climate Change: An Existential Threat to Corporations', in *Law Society of South Australia Bulletin*, 41/4 (2019): 28.
29. Joseph William Singer, *Entitlement: The Paradoxes of Property* (New Haven, CT: Yale University Press, 2000), 204 (emphasis added).
30. Duncan Kennedy, *Sexy Dressing Etc* (Cambridge, MA: Harvard University Press, 1993), 90–91 (emphasis in the original).

simple-minded sense that it *could* prohibit, but judges and legislators reject demands from those injured that the injurers be restrained.'[31] And they are invisible, in the sense '. . . that when lawmakers do nothing, they appear to have nothing to do with the outcome. But when one thinks that many other forms of injury are prohibited, it becomes clear that inaction is a policy, and that law is responsible for the outcome, at least in the abstract sense that the law "could have made it otherwise."'[32] Indeed, '[i]t is clear that lawmakers *could* require almost anything. When they require nothing, it looks as though the law is uninvolved in the situation, though the legal decision not to impose a duty is in another sense the cause of the outcome when one person is allowed to ignore another's plight.'[33]

This brings us full circle. The importance of relationship in understanding private property reveals an important, yet paradoxical, dimension of choice. It is simply this: the freedom that liberalism secures to the individual to choose a life project means that in the course of doing that, the individual also chooses the laws, relationships, communities, and so forth that constitute the political and legal order. In other words, in the province of politics people choose their contexts (through electing representatives, who enact laws and appoint judges who interpret those laws), which in turn defines the scope of one's rights—choice, decisionmaking authority—and the institutions that confer, protect and enforce it (bearing in mind the ground rules of permission as well as the ground rules of prohibition). Individuals as much choose the regulation of property as they do the control and use of goods and resources.[34]

How Private Property Facilitates the Externalities of Climate Change

When we focus on relationship as central to private property and the political-regulatory contexts we choose, we begin to see something

---

31. Kennedy, *Sexy Dressing Etc*, 91.
32. Kennedy, *Sexy Dressing Etc*, 91.
33. Kennedy, *Sexy Dressing Etc*, 91 (emphasis in the original, footnotes removed).
34. I am most grateful to Joseph William Singer for bringing this crucial point to my attention. And see also the essays collected in *Property and Community*, edited by Gregory S Alexander and Eduardo M Peñalver (Oxford, UK: Oxford University Press, 2010).

else that was always there, although it was hidden from our view. The externalities of private property create many other types of relationship in which the lives of many are controlled by the choices of a few.[35] Anthropogenic climate change is a stark example.

While the science is complex, it is clear enough that humans, through their choices, produce the gasses that enhance the natural greenhouse effect which heats the earth's surface.[36] Among other effects, anthropogenic climate change results in drought and desertification, increased extreme weather events, and the melting of polar ice (especially in the north) and so rising seas levels.[37] We might call this the 'climate change relationship'. And private property facilitates choice (both human and corporate) about the use of goods and resources in such a way that emits greenhouse gasses.

Our choices about goods and resources cover the gamut of our chosen life projects: where we live, what we do there, how we travel from place to place and so forth. Corporate choices are equally important, for they structure the range of choice available to individuals in setting their own agendas, thus giving corporations the power to broaden or restrict the meaning of private property in the hands of individuals. Green energy (solar or wind power), for instance, remains unavailable to the individual consumer if no corporate energy provider is willing to produce it.

Externalities do not end at the borders, physical or legal, of a good or resource; choices occur within a web of relationships, not only legal and social, but also physical and spatial. Who is affected? Everyone, the world over, with the poor and disadvantaged of the developing world disproportionately bearing the brunt of the human

---

35. On social-legal relationships, see William Twining, *General Jurisprudence: Understanding Law from a Global Perspective* (Cambridge, UK: Cambridge University Press, 2009), chapter 15, 1–7 (additional text) at <www.cambridge.org/twining>. Accessed 5 January 2021.

36. Lomborg, 'Introduction', 1.

37. Steve Lonergan, 'The Human Challenges of Climate Change', in *Hard Choices: Climate Change in Canada*, edited by Harold Coward et al (Waterloo, Canada: Wilfrid Laurier University Press, 2004), 13, 25, Figure 2.8. Schematic Diagram of Observed Variations, a) Temperature Indicators.

consequences of climate change[38]—decreasing security, shortages of food, increased health problems, and greater stress on available water supplies. Indeed, as Jedediah Purdy argues

> [c]limate change threatens to become, fairly literally, the externality that ate the world. The last two hundred years of economic growth have been not just a preference-satisfaction machine but an externality machine, churning out greenhouse gases that cost polluters nothing and disperse through the atmosphere to affect the whole globe.[39]

Consider human security. This will decrease both within countries affected directly by climate change, and in those indirectly affected through the movement of large numbers of people displaced by the direct effects of climate change in their own countries. In the case of rising sea levels, for instance, sixty percent of the human population lives within 100km of the ocean, with the majority in small- and medium-sized settlements on land no more than 5m above sea level. Even the modest sea level rises predicted for these places will result in a massive displacement of 'climate' or 'environmental refugees.'[40] And private property, through securing choice about the use of goods and resources to those in the developed world, makes all of this possible.

Conclusion

The theme of this special issue is, as Mike Hulme suggests, to ' . . . see how we can use the idea of climate change—the matrix of ecological functions, power relationships, cultural discourses and material flows that climate change reveals—to rethink how we take forward our political, social, economic and personal projects over the decades

---

38. Intergovernmental Panel on Climate Change (IPCC), *Climate Change 2007— Impacts, Adaptation and Vulnerability: Working Group II Contribution to the Fourth Assessment Report* (Cambridge, UK: Cambridge University Press, 2007), 7 at <http://www.ipcc.ch/ipccreports/ar4-wg2.htm>. Accessed 14 January 2010.

39. Jedediah Purdy, *A Tolerable Anarchy: Rebels, Reactionaries, and the Making of American Freedom* (New York, NY: Alfred A Knopf, 2009), 187. See also Jedediah Purdy, 'Climate Change and the Limits of the Possible', in *Duke Environmental Law & Policy Forum*, 18 (2008): 289.

40. Lonergan, 'The Human Challenges of Climate Change', 13, 25, Figure 2.8. Schematic Diagram of Observed Variations, a) Temperature Indicators.

to come.'[41] Before we pin our hopes on it as a cure-all, we might ask first whether the liberal concept of private property is ripe for just such a reappraisal. We can choose, but we must do so with our eyes open to the reality: that private property and the contexts in which we live are in fact *our* choice. Of course, government must act—the challenge of climate change is simply too large for individuals, even large numbers of them, to make a difference on their own. But make no mistake, individuals can and must choose a new way. In exercising choice about our context and about goods and resources, we must take responsibility, too.[42]

So, it is a matter of choice—not that of governments, but our own. I have previously written about this:

> So long as an individual, when faced directly with a clear and specific choice thinks first of themselves, free to choose to suit themselves, without any regard for others, then the consequences of anthropogenic climate change will inevitably follow. Unless property, or liberalism itself, is removed entirely, the state might control, and even prevent, *some* choices, but it cannot prevent *all* of them. This is further complicated by corporations: when those entities choose what to produce, then individual choice—about what to wear, sources of energy, and so on—is narrowed, and not always in a beneficial way.
>
> Property represents a tool, used by each of us, individually, and all of us, collectively, to choose how to use things. This in turn allows us to produce three climate consequences. First, spatially, we exert supreme, absolute, and uncontrollable power over the citizens of other nations, creating a set of unequal, or asymmetrical, relationships that alter the social, political and economic structures within those other nations. Second, temporally, we alter the social, political and economic structures of other nations for *future generations*. Third, and above all, through property we engage, perhaps unwittingly,

---

41. Mike Hulme, *Why We Disagree About Climate Change: Understanding Controversy, Inaction and Opportunity* (Cambridge, UK: Cambridge University Press, 2009), 362.
42. See James Hansen, *Storms of My Grandchildren: The Truth About the Coming Climate Catastrophe and Our Last Chance to Save Humanity* (New York, NY: Bloomsbury, 2009).

in what I have previously called 'eco-colonialism'; we colonise *both* other nations *and* our own; we are both coloniser and colonised.

The choice is ours: continue to use private property as a tool of eco-colonialism, or redefine its meaning so as to prevent climate consequences. The time to choose is now: the longer we wait, the closer we come to the catastrophe we are already beginning to live.[43]

Nothing has changed. The choice remains.

---

43. Paul Babie, 'Our Choice', in *Alternative Law Journal*, 44 (2019): 171.

Agathon: A Journal of Ethics and Value in the Modern World, Vol 8/2021

Emerging Perspectives on the Freedom to Choose: Two Pathways Out of Individualism

Jana L Norman

Abstract: Given the near total consensus (ninety-seven per cent) amongst climate scientists that climate-warming trends are extremely likely due to human activities, the significant percentages of people professing not to *believe in* anthropogenic climate change seems, in and of itself, incredible. This is the territory of 'psychology, sociology, anthropology', and the reality is that 'the debate over climate change, like almost all environmental issues, is a debate over culture, worldviews, and ideology'. The climate crisis is, in this sense, an existential crisis: 'to recognise greenhouse gases as a problem requires us to change a great deal of how we view the world and ourselves within it.' And no matter how illogical it seems in terms of the science, it makes sense that denial is one of the responses to this existential crisis. Denial is always part of human response to the upheaval, grief and loss we experience when our world changes either interpersonally or globally, as in the case of climate. A better story about who we are and our place in the world as human beings would go a long way towards helping us get our bearings in this reality, helping some move out of denial and ameliorating some of the adverse mental health effects for those of any political stripe who struggle with how to live in this reality and face its consequential challenges to our sense of identity, belonging and purpose in the world. This article presents two projects—one from a Christian perspective and one from the perspective of the philosophy of science—which can be characterised as the twin pillars of the western social imaginary. These stories neither diminish the self nor limit choice, but offer a corrective to individualism, reframing the value of freedom as the freedom to choose life.

Introduction

In his 2012 article, 'Climate science as culture war', and subsequent 2015 book, *How Culture Shapes the Climate Change Debate*, business and environment professor Andrew J Hoffman analyses the partisan divide in beliefs in the United States about the facts and causes of climate change. In the article, Hoffman cites a decline in the number of conservatives and Republicans who believe that global warming has already begun (from fifty per cent to thirty per cent between 2001 and 2010) and an increase with reference to the same in the same period amongst liberals and Democrats (from sixty per cent to seventy per cent).[1] If Hoffman were writing the article today, he could note that, although more Americans overall say that climate change is a major threat to the wellbeing of the United States, the partisan divide still stands: findings of a survey conducted by the Pew Research Center in March 2020 indicate that the percentage of Democrats who feel climate change is a threat has increased by seventy per cent since 2009 (up to 88 percent) and that the 6 percent increase amongst Republicans who feel similarly is not statistically significant (with the total being just thirty-one per cent).[2] Beliefs about climate change fall similarly across partisan divides in Australia. In February 2020, the *Sydney Morning Herald* reported results from two focus groups held for their publication and *The Age*, indicating a 'highest ever' percentage of Liberal and Nationals voters reporting 'serious doubts' about whether climate change is occurring, compared to a 'lowest ever' percentage of doubt in this regard for Labor voters.[3]

Given the near total consensus (ninety-seven per cent) amongst climate scientists that '[c]limate-warming trends are extremely likely due to human activities',[4] the significant percentages of people professing not to *believe in* anthropogenic climate change seems, in and

---

1. Andrew J Hoffman, 'Climate Science as Culture War', in *Stanford Social Innovation Review* (2012): 30.
2. Brian Kennedy, 'U.S. Concern About Climate Change is Rising, But Mainly Among Democrats', in *Pew Research Center Fact Tank* (16 April 2020) at <https://www.pewresearch.org/fact-tank/2020/04/16/u-s-concern-about-climate-change-is-rising-but-mainly-among-democrats/>. Accessed 5 January 2021.
3. David Crowe, 'Australians Back Climate Change Action While Science Divides Along Party Lines' in *Sydney Morning Herald* (4 February 2020) at <https://www.smh.com.au/politics/federal/australians-back-climate-change-action-while-science-divides-along-party-lines-20200203-p53x7y.html>. Accessed 5 January 2021.
4. 'Scientific Consensus: Earth's Climate is Warming', in *NASA Global Climate Change: Vital Signs of the Planet* at <https://climate.nasa.gov/scientific-consensus/>. Accessed 5 January 2021.

of itself, incredible. Apparently, facts alone are not persuasive in matters so confronting and so far beyond, in complexity, most people's capacity for processing scientific information.[5] Hoffman points out that, whilst '[p]hysical scientists may set the parameters for understanding the technical aspects of the climate debate . . . they do not have the final word on whether society accepts or even understands their conclusions'.[6] This is the territory of 'psychology, sociology, anthropology',[7] and the reality is that 'the debate over climate change, like almost all environmental issues, is a debate over culture, worldviews, and ideology'.[8] The climate crisis is, in this sense, an existential crisis: 'to recognise greenhouse gases as a problem requires us to change a great deal of how we view the world and ourselves within it'.[9]

No matter how illogical it seems in terms of the science, it makes sense that denial is one of the responses to this existential crisis. Denial is always part of human response to the upheaval, grief and loss we experience when our world changes either interpersonally or globally, as in the case of climate. It also makes sense that denial manifests itself so vociferously at the conservative end of the political spectrum. Anthropogenic climate change poses a very specific threat to the nub of the traditional western worldview. One cannot believe in strident individualism and the absolute value of freedom of choice whilst also admitting the deep entanglements that the reality of our current global environmental situation reveals. Insofar as individualism is a function of the denial of dependency and interdependency of the self upon and with others, what anthropogenic climate change exposes is the truth of our total and fundamental dependency on others, both human and non-human. It occurs to me that, for some, the denial of anthropogenic climate change is a doubling down on that other, more fundamental denial on which western ideology is based.

---

5. See Hoffman, 'Climate Science as Culture War', 32: 'With upwards of two-thirds of Americans not clearly understanding science or the scientific processes and fewer able to pass even a basic scientific literacy test, according to a 2009 California Academy of Sciences survey, we are left to wonder: How do people interpret and validate the opinions of the scientific community?'
6. Hoffman, 'Climate Science as Culture War', 32.
7. Hoffman, 'Climate Science as Culture War', 32.
8. Hoffman, 'Climate Science as Culture War', 32.
9. Hoffman, 'Climate Science as Culture War', 33.

Even for those who are not doubling down on denial, these are existentially challenging times. It is my contention that a better story about who we are and our place in the world as human beings would go a long way towards helping us get our bearings in this reality, helping some move out of denial and ameliorating some of the adverse mental health effects[10] for those of any political stripe who struggle with how to live in this reality and face its consequential challenges to our sense of identity, belonging and purpose in the world. If we, humans, are the problem, then we need a better story of ourselves and our place in the material world to lead us out towards becoming part of whatever solutions may yet be possible. As Hoffman points out in discussing techniques for generating consensus-based discussion around climate change, an important aspect of the attempt to come to terms with these world and worldview altering issues is to be invited to engage with them from within the right 'broker frames':

> People interpret information by fitting it to pre-existing narratives or issue categories that mesh with their worldview. Therefore, information must be presented in a form that fits those templates, using carefully researched metaphors, allusions, and examples that trigger a new way of thinking about the personal relevance of climate change. To be effective, climate communicators must use the language of the cultural community they are engaging.[11]

This is what I contend that the two projects explored in this article accomplish: careful research, from within two different issue categories, that triggers a new way of thinking about the personal relevance of climate change. I do not suggest that the two projects, one from a Christian perspective and one from the perspective of the philosophy of science, are either opposing or complementary. I simply take note

---

10. For an introductory overview of current issues in climate emotions such as fear and anxiety, see Christine Ro, 'The Harm from Worrying About Climate Change', in *BBC Future* (10 October 2019) at <https://www.bbc.com/future/article/20191010-how-to-beat-anxiety-about-climate-change-and-eco-awareness>. Accessed 5 January 2021. For a more comprehensive exploration of the topic, cited in the Ro article, see S Clayton et al, *Mental Health and Our Changing Climate: Impacts, Implications, and Guidance* (Washington, DC: American Psychological Association and ecoAmerica, 2017).

11. Hoffman, 'Climate Science as Culture War', 35.

that, since these two cultural sub-communities of Christianity and science can be characterised, arguably, as the twin pillars of the western social imaginary,[12] new stories emerging from within and around them may be quite accessible for those with ears to hear their different dialects. In this article, I briefly describe each project and comment on what they offer their audiences by way of existential guidance for our times.

Project 1: Recovering a Biblical Voice About Owning and Consuming

Paul Babie and Michael Trainor, in their book *Neoliberalism and the Biblical Voice: Owning and Consuming*, examine the original socio-historical contexts out of which the Christian scriptures emerged in order to surface a new story about the meaning of freedom of choice in our lives. In the contemporary story of neoliberalism that animates the western cultural paradigm (ubiquitous and dominating in its global reach), the freedom to choose is an ultimate good in and of itself. As these authors note, however, in the context of anthropogenic climate change driven by consumer choices for goods, the extraction, manufacture and transportation of which generate greenhouse gases, we are confronted with the reality that the raw value of choice is necessarily tempered by the consequences of its exercise:

> Choice, in the modern, neoliberal world, then, is both a gift and a constraint: a gift in the sense that we are free to choose so as to suit our own, individualised preferences; but also a constraint because in making those choices, we foreclose other possibilities, both for ourselves and for others.[13]

At the heart of Babie and Trainor's project is the quest to fill 'a hole or gap which emerges from liberalism itself, and which is seen more clearly in the case of private property'.[14] Babie and Trainor define this gap as the absence in neoliberalism of anything to guide our choices

---

12. The genealogy of western dualism depicted in Val Plumwood, *Feminism and the Mastery of Nature* (Abingdon, UK: Routledge, 1993) is instructive on this point.
13. Paul Babie and Michael Trainor, *Neoliberalism and the Biblical Voice: Owning and Consuming* (Abingdon, UK: Routledge, 2018), 134.
14. Babie and Trainor, *Neoliberalism and the Biblical Voice: Owning and Consuming*, 134.

around consumerism and ownership as these choices relate to others or to the common good. They observe that this ideology begins and ends at the espoused ultimate value of choice and offers nothing beyond that assertion in terms of 'what is a good or a right or a just choice'.[15] As researchers in the area of intersections between religion and contemporary issues, including law and the environment, their quest to fill this void involves working to 'recover and re-engage an ancient wisdom articulated in the teachings and deeds of the Galilean Jesus and by the writers of Mark, Q and Luke who were inspired by the practice of Jesus'.[16] This recovery process revolves around putting these writings in their historic socio-economic context and linking them to 'the economic vision of the Galilean Jesus concerned (with) the appropriate use of property and possessions for the good of others and the cohesion of village life'.[17] The work of critical biblical exegesis yields a number of core themes about choice in relation to possessions and property, which Babie and Trainor transpose for the present into a set of seven theses that they believe have 'implications for our engagement with the neoliberal perspective on private property and choice'.[18]

Here is my reading of the new story about how to view the world and our place in it that Babie and Trainor surface for a critically astute Christian audience. As the authors explicate in a lengthy discussion of the meaning of private property, property is actually a relation between people: those who own this or that thing and those (both universally and specifically *visa vie* contract) who do not own that same thing. For a Christian, the meaning of the property relation is located within the context of relationship with God. The directionality of this aspect of relationship with God is not, as has been part of an old religious story in both ancient and modern times, that property

---

15. Babie and Trainor, *Neoliberalism and the Biblical Voice: Owning and Consuming*, 134

16. Babie and Michael Trainor, *Neoliberalism and the Biblical Voice: Owning and Consuming*, 135.

17. Babie and Michael Trainor, *Neoliberalism and the Biblical Voice: Owning and Consuming*, 99.

18. Babie and Michael Trainor, *Neoliberalism and the Biblical Voice: Owning and Consuming*, 136.

and possessions are demonstrations of God's favour.[19] Rather, faith and trust in God liberate believers from anxiety and acquisitiveness with regard to possessions which, in turn, frees them to participate in community with generosity and kindness. Possessions—food, property, wealth—and the status they confer—are not rewards for faithfulness; the sharing of possessions is an expression of faithfulness. Such perspective engenders a gracious economy in which needs are met. This mode of participation in human community in which self-interest is subsumed within the greater good is expressed as God's intention for God's people through the Torah. Babie and Trainor argue that this 'desire to renew the heart of the Torah' fuelled the 'economic vision of the Galilean Jesus', and that 'the spirit of this vision from the historical Jesus' was 'absorbed' by Mark and the other writers of the first century CE Jesus movement, who 'reshaped it to speak pertinently to Jesus followers and gospel's addressees a generation or two later'.[20]

Babie and Trainor's project, then, can be characterised as another reshaping of the source material associated with the Galilean Jesus, refracted through the first generations of his followers, for a new generation of followers. They acknowledge the 'hermeneutical leap'[21] involved in bringing ancient texts so far forward in time, but they build a bridge to span the distance by drawing clear connections between ancient and contemporary economic realities. Babie and Trainor dedicate chapters to describing the Roman-Mediterranean market place on the one hand, and 'how private property acts as the deployment of neoliberal rights as part of the neoliberal rationality of our world'[22] today, on the other. The growing disparity between the poor and the powerful in ancient Rome and Jerusalem, with the associated pressures to conform to dehu-

---

19. See Babie and Michael Trainor, *Neoliberalism and the Biblical Voice: Owning and Consuming*, 108: In discussion about the focus of Mark 10.23-27 on how one gains entry into the reign of God, Babie and Trainor identify the roots of what is known today at the 'prosperity gospel': an attitude in first century Christianity that reflects the point of view 'that wealth is the fruit of God's predilection for some who appear to be more favoured than the poor'.

20. Babie and Trainor, *Neoliberalism and the Biblical Voice: Owning and Consuming*, 99.

21. Babie and Michael Trainor, *Neoliberalism and the Biblical Voice: Owning and Consuming*, 34.

22. Babie and Michael Trainor, *Neoliberalism and the Biblical Voice: Owning and Consuming*, 71.

manising and fragmenting market forces, speaks into the situations of people trying to navigate against these same types of forces in different garb today. As Babie and Trainor note, in the first century Roman imperial world:

> a small number of elite ruled the majority of non-elites as they exercised power in a vertical axis—with elites on the Empire's periphery bonded closely with those at the imperial centre, in Rome and other major cities—especially through taxation. Overall, the small aristocratic powerful group consumed approximately 65 percent of production.[23] The hierarchical and vertical axes of Roman society favoured access to material resources by the social elite and wealthy. They also determined the burden of responsibility for the production of enough goods and foodstuffs from the majority of peasants to feed and materially support them.[24]

This picture sits alongside the statistic in a January 2019 report produced by Oxfam International, that in 2018 the 26 richest people in the world held as much wealth as half of the global population (the 3.8 billion poorest people).[25] Babie and Trainor point to a tenant of economic theory that links ancient times to modern: 'whatever system of allocation a social system might choose to use, there are going to be some people who win and some who lose. Some will get some of the stuff, whatever it is, about which there is competition, and others will not'.[26] Accompanying this pan-temporal economic dynamic, they argue (thus forming the purpose of their project), is the possibility that moral agency may be expressed counter-culturally via a com-

---

23. Warren Carter, *The Roman Empire and the New Testament: An Essential Guide* (Nashville, TN: Abingdon Press, 2006), 3, cited in Babie and Trainor, *Neoliberalism and the Biblical Voice: Owning and Consuming*, 47.
24. Babie and Trainor, *Neoliberalism and the Biblical Voice: Owning and Consuming*, 47.
25. Oxfam, *Public Good or Private Wealth? Oxfam Briefing Paper* (January 2019), cited in *World Social Report 2020: Inequality in a Rapidly Changing World* (New York, NY: United Nations Department of Economic and Social Affairs, 2019), 31, n 26, with the caveat that because '[w]ealth is particularly challenging to estimate in poor countries and for people that have negative wealth (debt and mortgages, for example)' this type of information 'should be interpreted with caution'.
26. Babie and Trainor, *Neoliberalism and the Biblical Voice: Owning and Consuming*, 71.

mitment to the common good, as per and insofar as one is able given one's position and capacity.[27]

I leave it to the reader of Babie and Trainor's book to experience the specifics of their investigation, which is quite intricate in its detail, comprising socio-economic contextualisation, etymology of the Greek, and structural literary exegesis. What suits my purpose in this article is to highlight the result of their efforts: a framework to assist a particular audience in dealing productively with the existential challenge of anthropogenic climate change. There are pathways towards new loci of meaning in what Babie and Trainor extract from sacred texts, specific for those contemporary Christians who may be asking, 'What would Jesus do?' The framework (that is a pathway) suggested by Babie and Trainor consists of one overarching principle and an array of themes which can guide the exercise of choice with regard to private property and material possessions. In terms of the overarching principle, Babie and Trainor suggest that 'the ancient view prioritised the Other and the community, the social well-being'.[28] Aligning one's choices with this priority, for contemporary Christians, is a matter of choosing: the sacred; wholeness; a 'way of living that is open to others'[29]; for the poor; for generosity; and for friendship, or the proactive overcoming of social divisions.[30] Although skeletal at the stage of development which Babie and Trainor accomplish in their book, as they admit, this framework already gives more guidance than is otherwise available in contemporary neoliberalism for how to enact the power of choice this ideology enshrines. Babie and Trainor conclude that the theological voice they trace back to primary sources 'points us towards the Other and calls us to use the power we have been given in choice so as to benefit all rather than the few'.[31]

---

27. Babie and Trainor choose Mark, Q and Luke as their primary sources, noting the probable differences in the audiences for these various writings, which range from peasant villagers to an urban audience that may have included some from upper echelons of society. See Babie and Trainor, *Neoliberalism and the Biblical Voice: Owning and Consuming*, 112–113, for references to these different audiences with regard to Q and Luke, for example.
28. Babie and Trainor, *Neoliberalism and the Biblical Voice: Owning and Consuming*, 135.
29. Babie and Michael Trainor, *Neoliberalism and the Biblical Voice: Owning and Consuming*, 139.
30. Babie and Michael Trainor, *Neoliberalism and the Biblical Voice: Owning and Consuming*, 141.
31. Babie and Michael Trainor, *Neoliberalism and the Biblical Voice: Owning and Consuming*, 143.

Project 2: Meeting the Universe Halfway

The process of developing a productive story that might guide a particular community through the turbulence of the existential crisis evoked by the anthropogenic climate crisis, as I have outlined it in the project described above, begins with a focus on first principles and primary sources. For critical theologians like Babie and Trainor, this means extracting themes from the earliest available Christian writings, which they then translate into a framework of ethical guidance for their particular critically reflective Christian audience. For theoretical particle physicist and philosopher of science Karen Barad, the primary source is the smallest observable matter of the universe, or what could be considered the most primary or fundamental aspect of physical existence. Barad is able to observe, describe and discuss the 'basic unit of reality on the level of empirically verifiable properties of matter'.[32] From her empirical observations of matter, Barad constructs, for her particular audience, a framework that serves as a pathway or guidance for reimagining what it means to be a human being in our times.

The audience for the story arising from quantum physics is not as easy to collect under one descriptor as an audience for a biblical voice would seem to be. Although there is all manner of variation on the theme 'Christian', this is an existing category related to a particular religious frame of reference. How to categorise and characterise the 'quantum crowd'? Perhaps only by distinction. The distinction I would draw in contrast to a theological audience relates to the horizon of reference engaged by each: for the theologically oriented it is transcendent, whereas for this audience it is immanent. Making this distinction does not preclude any theologically oriented persons from the possibility of finding meaning in and responding to a story that arises out of scientific observation of the natural world rather than spiritual resources, or even vice versa. It is not necessary to presume that the audiences are mutually exclusive; just distinctive.

In a principal text on her process and findings, *Meeting the Universe Halfway: Quantum physics and the entanglement of matter and meaning*, Barad describes her work as experimenting at the

---

32. Sherryl Vint, 'Entangled Posthumanism: Review of *Meeting the Universe Halfway: Quantum Physics and the Entanglement of Matter and Meaning* by Karen Barad', in 35/2 *Science Fiction Studies* (2008): 313, 315.

metaphysical level. What she is referring to is that the nature of the work—observing the sub-atomic level of material existence—reveals a paradox, the theoretical resolution of which carries profound onto-logical, epistemological and ethical implications: why is it possible that in some instances matter appears to behave like a particle and in others like a wave? Confronting this paradox led twentieth century Danish physicist Niels Bohr to theorise the indeterminacy of matter. It is the observing itself that determines which way matter presents itself, which is to say: how it materialises.

Following on from Bohr,[33] Barad puts resolution of the appar-ent paradox this way: the primary ontological reality—the nature of things—is phenomena, not entities. Matter and meaning emerge together in what she calls intra-acting. Not interacting between two pre-determined objects, but intra-acting; co-worlding. Experiments do not reveal the 'pre-existing determinate nature of the entity being measured';[34] they reveal intra-action or the co-emergence of matter and meaning. For Barad, the term 'intra-action' usefully describes 'ontological inseparability, in contrast to the usual 'interaction,' which relies on a metaphysics of individualism (in particular, the prior exis-tence of separately determinate entities)'.[35] Ontological inseparability at the sub-atomic level means that we are already a part of everything, as demonstrated by our attempts to observe something as Other, as if we are not part of it. This theory of indeterminacy, which Barad calls agential realism, reveals that there can be no actual other because we are *always already* there.

---

33. Vint, 'Entangled Posthumanism', 319: 'Barad's certainty that Bohr is correct and that his model of indeterminacy is the best way to conceive of the physical universe relates in part to recent changes in experimental physics. Previously, including during Bohr's lifetime, most quantum physics experiments were conducted via conceptual experimental metaphysics, used to investigate through philosophical debate questions that could not be tested in empirical reality. Such experiments, called gedanken experiments in physics, are impractical to carry out but nonetheless useful for the insights they generate when reasoned through. Many of Bohr's gedanken experiments included the design for apparatuses to test his theories, but physically constructing such apparatuses was not within reach until quite recently, when Bohr's theories were borne out by empirical testing.'

34. Vint, 'Entangled Posthumanism', 315.

35. Vint, 'Entangled Posthumanism', 315.

There is a profound ethical implication to this insight about the nature of existence, beautifully described by Barad:

> A delicate tissue of ethicality runs through the marrow of being. There is no getting away from ethics—mattering is an integral part of the ontology of the world in its dynamic presencing. Not even a moment exists on its own. 'This' and 'that', 'here' and 'now', don't pre-exist what happens but come alive with each meeting. The world and its possibilities for becoming are re-made with each moment. If we hold onto the belief that the world is made of individual entities, it is hard to see how even our best, most well-intentioned calculations for right action can avoid tearing holes in the delicate tissue structure of entanglements that the lifeblood of the world runs through. Intra-acting responsibly as part of the world means taking account of the entangled phenomena that are intrinsic to the world's vitality and being responsive to the possibilities that might help us and it flourish. Meeting each moment, being alive to the possibilities of becoming, is an ethical call, an invitation that is written into the very matter of all being and becoming. We need to meet the universe halfway, to take responsibility for the role that we play in the world's differential becoming.[36]

The Barad text is a tome of over 500 pages and is commendable for its carefulness and accessibility. As one reviewer notes, '(Barad) provides an excellent overview of science studies . . . (s)he also meticulously takes the reader through a number of physics experiments, demonstrating that Bohr's model of complementarity is thus far the most accurate description we have of the nature of the universe.'[37] Reproducing the details of the science behind the story of intra-action Barad tells is beyond my purpose here. My purpose is to point out that the story offers another framework for finding a meaningful way to be present and purposeful in our times:

(Barad) calls upon us to take responsibility for 'the possibilities for what the world may become' that she reminds us are continually open to us simply through our way of intra-acting, of bringing forth the world in each moment.[38]

---

36. Karen Barad, *Meeting the Universe Halfway: Quantum Physics and the Entanglement of Matter and Meaning* (Durham, NC: Duke University Press, 2007), 396.
37. Vint, 'Entangled Posthumanism', 315.
38. Vint, 'Entangled Posthumanism', 318.

Conclusion

In 1993, critical ecological feminist Val Plumwood published a comprehensive analysis of the logic of dualism that underpins western philosophy and culture, noting the key role in this ideology played by the naturalised hierarchy of humans over nature (justified by the presumption that reason is a superior capacity and is exclusive to humans). This presumption of separateness and assignation of superiority is further iterated, argues Plumwood, across various 'types' or 'categories' of humans so that difference becomes a power dynamic not only in human-earth relations but also in human relations. The powers of reason are ascribed only to certain humans (white, male, elite), lumping all that are not these particular humans on the underside with nature.[39] The effect of dualism as an 'alienated form of differentiation'[40] is experienced as a process of Other-ing, whereby the underside of the hierarchy is defined in terms of lack by comparison to the upperside. This lack justifies an instrumental orientation on the part of the upperside towards the underside, which in turn justifies exploitation and oppression of the Other, both human and non-human.

One of the most striking features of dualism which Plumwood exposes is its foundation in denial, fundamentally the denial of the dependency of the upperside on the underside which is, finally, a denial of the interdependency of all life. This denial takes its most absurd form in the context of the human-earth relationship: in what way can humans ever be independent of the earth? This foundational (to western ideology) denial of dependency and interdependency

---

39. See Plumwood, *Feminism and the Mastery of Nature*, 43–44, for an illustrative list of the 'set of interrelated and mutually reinforcing dualisms which permeate western culture' that hang off the central reason/nature divide, encompassing various forms of difference associated with concepts such as gender and race, and a discussion of the key role gender plays in 'the exclusions of reason': 'The exclusions of reason are multiple and not reducible to those of gender. Nevertheless, gender plays a key role, since gender ideals especially involve ideals of reason [GE Lloyd, *The Man of Reason* (London, UK: Methuen, 1984)], and male ideals which lay claim to universality for men often invoke the elite male identity of the master. Thus, to read down the first side of the list of dualisms is to read a list of qualities traditionally appropriated to men and to the human, while the second side present qualities traditionally excluded from male ideals and associated with women, the sex defined by exclusion ...'

40. Plumwood, *Feminism and the Mastery of Nature*, 42.

also permeates human relations in this culture, where individualism is lauded as an ultimate value. As the world reaps the consequences of this ill-logic, it is no wonder we of the western world experience these consequences as an existential crisis.

It was reading Plumwood that put me onto the quest for better stories that may help people of the western cultural imaginary by serving as guide in this crisis. Plumwood herself describes what she calls an 'escape route' from dualism, but she could see no story compelling enough to set us on that course. Back in 1993, in *Feminism and the Mastery of Nature*, Plumwood wrote that she could see no story of human significance emerging that would helpfully replace the traditional religious story that had started showing cracks in the Enlightenment and was finally toppled by the Modernist response of rationalism to the existential crises evoked by two world wars: the death of God. Plumwood describes the resultant dearth of meaning in two complementary statements near the end of the book:

> No single position on human significance has appeared to replace that of otherworldly religion; rather there are a number of sons contending for the mantle of the Father, the power to confer meaning and identity.[41]

> Contemporary western identity has rejected the otherworldly significance and basis for continuity, but has given it no other definitive meaning, provided no other satisfactory context of continuity or embeddedness for human life.[42]

My thesis is that new stories are now emerging from within western cultural consciousness, like mushrooms popping up in the decayed matter of the old worldview. I celebrate the two stories featured in this article for their potential as freshly satisfying contexts of continuity and embeddedness, each for their different audiences. No grand narratives here; just small lights to guide the way. The one: a this-worldly conferral of religious meaning and identity in a renewed vision of being the people of God on earth, in which generosity and friendship guide participation in community as people make choices about their use of and relation to private property. The other: a striking recast of

---

41. Plumwood, *Feminism and the Mastery of Nature*, 101.
42. Plumwood, *Feminism and the Mastery of Nature*, 101.

material identity as entanglement in which indeterminacy opens up space for co-worlding with the Other/non-Other as a framework for the necessary consumption that marks our dependent earth existence. I can make no predictions about the potential take-up of these stories or whether they will prove to soothe souls and transform behaviours, personal and political. But I am hopeful because, for those ready to let go of denial about the reality of anthropogenic climate change, these stories lay positive pathways out of the other, more fundamental denial at the root of western culture. These stories neither diminish the self nor limit choice even as they offer an urgently required corrective to individualism, reframing the value of freedom as the freedom to choose life as it well and truly is according to both science and religion: interdependent and entangled.

Contributors

Bios:

Paul Babie is ALS Professor of Property Law, Associate Dean of Law (Learning & Teaching), and Director of the Law and Religion Project of the Research Unit for the Study of Society, Ethics and Law (RUSSEL) (which he founded in 2007). He was elected a Fellow of the Australian Academy of Law in 2017. His research, throughout his career, has involved asking what property is and how, if at all, it can be justified. He has explored those questions from legal theoretical and from theological perspectives. He teaches property law, property theory, law and religion, and Roman law.

Hugh Breakey is a Senior Research Fellow in moral philosophy at Griffith University's Institute for Ethics, Governance & Law.

Donald A Brown is Scholar in Residence on Sustainability Ethics and Law at Widener University School of Law.

Peter Burdon is Associate Professor at the Adelaide Law School.

Seforosa Carroll is an ordained minister of the Uniting Church in Australia. She is an Australian-Rotuman, originally from Fiji. Sef is currently Programme Executive for Mission from the Margins/ Ecumenical Indigenous Peoples Network, World Council of Churches, Geneva. Prior to this, Sef worked with UnitingWorld in Australia as a theological researcher (climate and gender) and church partnerships manager for the Pacific. Sef is a research fellow with the Center for Public and Contextual Theology (PaCT), Charles Sturt University, Australia and a member of the Center of Theological Inquiry, Princeton.

Brendan Mackey is Director of the Griffith Climate Action Beacon, Griffith Climate Change Response Program, and the National Climate Change Adaptation Research Facility at Griffith University.

Jana Norman is a researcher in the Faculty of Arts at the University of Adelaide with a focus on posthuman constructs of human identity, following the completion of doctoral research in legal theory and Earth Jurisprudence at the Adelaide Law School. Her book, Posthuman legal subjectivity: reimagining the human in the Anthropocene, is forthcoming from Routledge in the series 'Law, Justice and Ecology'.

Prue Taylor is a senior lecturer at the University of Auckland, School of Architecture and Planning.

Claire Williams is a climate scientist and practicing human rights lawyer with a particular focus on Aboriginal heritage and Native Title issues. She is currently completing a PhD at Adelaide University which looks at the disconnect between scientific knowledge of human impact on Earth and the laws and policies which govern human behaviour. Claire is also interested in researching how long-term trends in large scale climate drivers affect local weather patterns. Claire received the Chancellor's Letter of Commendation for Academic Excellence for her work in ocean atmosphere interactions at Flinders University and has been a recent recipient of the Endeavour Scholarship (formally the Prime Minister's Award).

CPSIA information can be obtained
at www.ICGtesting.com
Printed in the USA
JSHW021516150723
44805JS00002B/129